T0325112

ENGINEERING IN TIME

The Systematics of Engineering
History and its Contemporary Context

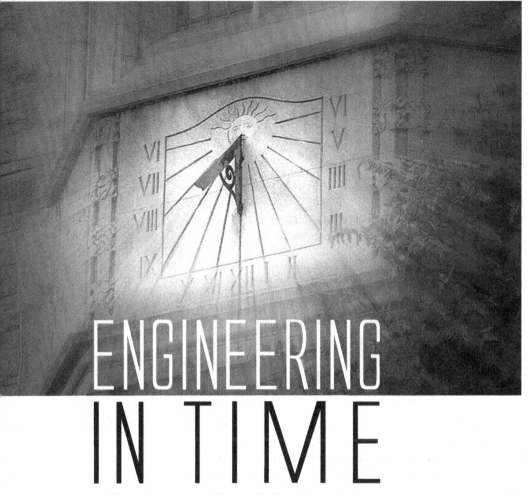

ENGINEERING
IN TIME

The Systematics of Engineering
History and its Contemporary Context

A A Harms
McMaster University

B W Baetz
Tulane University

R R Volti
Pitzer College

Imperial College Press

Published by

Imperial College Press
57 Shelton Street
Covent Garden
London WC2H 9HE

Distributed by

World Scientific Publishing Co. Pte. Ltd.
5 Toh Tuck Link, Singapore 596224
USA office: 27 Warren Street, Suite 401-402, Hackensack, NJ 07601
UK office: 57 Shelton Street, Covent Garden, London WC2H 9HE

British Library Cataloguing-in-Publication Data
A catalogue record for this book is available from the British Library.

First published 2004
Reprinted 2005, 2008

ENGINEERING IN TIME
The Systematics of Engineering History and Its Contemporary Context

ISBN-13 978-1-86094-433-8
ISBN-10 1-86094-433-7

Editor: Tjan Kwang Wei

Typeset by Stallion Press
Email: enquiries@stallionpress.com

Printed in Singapore

Preface

The history of human events is characterized by a vast range of thought and actions. Among those of particular consequences are the innumerable *ingenious devices* which have been conceived and produced to serve a variety of human interests. It is this basic and enduring human endeavor which identifies the history and contemporary context of engineering.

In its earliest form, engineering involved the making of stone tools and other artifacts to aid in human survival. During the ensuing millennia, the manufacture of ingenious devices expanded and contributed to the shaping of civilizations, to the establishment of human institutions, and to the enhancement of standards of living. Now, in the 21st century, engineering may be viewed as a profession which involves creative thought and skilled actions related to conceptualizing, planning, designing, developing, making, testing, implementing, using, improving, and disposing of a variety of devices, invariably seeking to meet a perceived societal interest. In these various functions, engineers connect the natural world of materials and phenomena with the societal world of needs and aspirations.

This function of connecting nature and society has traditionally been dominated by the premise that — from an engineering point of view — nature could be viewed as both an immense material resource and a vast depository for discards. But an expanded perspective has emerged, prompted by a wider recognition that the natural world is becoming a threatened meld. To be sure, educators, accreditation boards, and professional associations have long challenged engineers to adopt a broader intellectual versatility in the design of devices and to acquire a deeper sensitivity to the range of their impact. The proposition pursued here is that such a more concordant orientation requires specifically an integrated nature-engineering-society perspective — a perspective well served by a systematic view of historical chronology and recognition of an interconnected contemporary context of engineering. It is in this process of blending that a much more coherent and informing characterization of engineering can be established.

In order to develop such an engineering characterization, it is important that an overarching sweep of the engineering past, sequentially characterized in terms of selected pace-setting innovations and evolving connections, be related to aspects of the present and projected expectations of relevance to the theory and practice of engineering. For this purpose, selected aspects of the theory and practice of engineering-in-time are developed by an emphasis on three themes:

Part A: Introduction to Engineering, Chapter 1,
Part B: History of Engineering, Chapters 2 to 8,
Part C: Contemporary Context of Engineering, Chapters 9 to 13.

Part A introduces *Engineering in Time* by laying some conceptual and analytical foundations essential to the subsequent development. The idea of a *progression* and its notational incorporation is central for this purpose. A critical definition of engineering, together with two corollaries, is also introduced.

Part B contains two superimposed histories: one is a chronology of engineering invention and innovation from prehistory to contemporary times and the other is a simultaneously evolving interactive connectivity with engineering as a critical functional component. Thus, a traditional engineering history is integrated with its intrinsic and adaptive linkage processes.

Part C provides a contemporary context for an exposition of topics of special relevance to engineering. Among the subjects covered are natural phenomena and dynamics, professionalism and ethics, invention and innovation, device reliability and failure, risk and safety, public interaction, market penetration dynamics, material flow metabolism, and other related issues. Explicit mathematical formulations are selectively introduced and solutions discussed in terms of the graphical interpretation of differential equations.

As a means of insuring both clarity and unity in the development of these themes, the fundamental role of invention and innovation are embedded in informative heterogeneous progressions and illustrative adaptive networks. Then, major emphasis is placed on two principles: a general and systematic exposition is chosen rather than one which is encyclopaedic in case-study detail, and symbolic and graphical means are employed as important pedagogical tools. A decade of teaching using this format has shown that these features are most effective in depicting the evolving history of engineering as a complex network-based progression while simultaneously highlighting the dynamics of themes of relevance to engineering.

Archie Harms   Brian Baetz   Rudi Volti
McMaster University   Tulane University   Pitzer College
Hamilton, ON   New Orleans, LA   Claremont, CA

January 2004

Instructional Format

Our instructional experience using various earlier versions of this text suggests its use as basic and engineering accreditation related in two distinct undergraduate classroom settings:

(a) The entire text is used in a comprehensive 3 hour/week semester course at the upper division level, or

(b) Part A and Part B are used for a 2 hour/week course at the lower division level with Part C then forming the basis for a subsequent senior discipline-specific seminar course.

Additionally, we have found that this text to be effective in directing and focusing student efforts in support of term papers or group projects which have a historical-developmental component.

Acknowledgment

This book has a long and absorbing history. It began over two decades ago when one of us (AAH) became involved in the arduous process of undergraduate engineering program accreditation and the associated expectations for a complementary instructional component in the humanities and social sciences. This experience triggered a trek punctuated by discussions with accreditation committee members, industrial colleagues, and engineering program directors, eventually leading to lecture notes, student handouts, draft chapters, conference presentations, and the merging of academic-professional traditions of the authors. At various times along the way, specialized topics in this text have had the benefit of discussion with a number of university colleagues: Robert Baber, Douglas Davidson, William Harris, Hans Heinig, Gordon Irons, Howard Jones, Reuven Kitai, Patrick Nicholson, Gunhard Oravas, Les Shemilt, Byron Spencer, John Vlachopoulos, and Dave Weaver. Specific comments from off-campus colleagues include Bruno Augenstein (RAND Corp.), Doug Barber (Gennum Corp.), Gordon Dey (Mitel Corp.), Tom Fahidy (University of Waterloo), Eric Klaassen (Burlington Technologies), Gene Moriarty (San Jose State University), Klaus Schoepf (University of Innsbruck), and Jim Tiessen (Sunoco Corp.). Additionally, suggestions have been provided by Paul Challen, Theodore Harms, Nicholas Leeson, Robert Nau, Jonathan Stolle, Sharon Whittle, and numerous engineering students who over the years made perceptive comments on various drafts. Lastly, some of the concepts developed here have previously found their way into conference presentations and refereed journals thereby yielding useful suggestions and confirmations from interested listeners and anonymous reviewers. Word processing has been provided by Janet Delsey and Laura Honda and art work has been prepared by Eugene Martinello.

To all who have contributed in various ways, a sincere thank you.

Contents

10. Engineering: Patterns and Specializations 209

$N(t) \rightarrow E(t) \rightarrow D(t) \cdots$

11. Devices: Properties and Functions 243

$\cdots \rightarrow E(t) \rightarrow D(t) \rightarrow S(t) \rightarrow \cdots$

12. Society: Involvement and Ramifications 267

$\cdots \rightarrow D(t) \rightarrow S(t) \rightarrow R(t)$

APPENDICES

PART A

Introduction to Engineering

Engineering was initiated by the human imagination to serve a human purpose. The objective in this introductory Part A is to identify a framework for this human activity as an aid in thinking about engineering and its implications.

CHAPTER 1

About Engineering

Identifying a Framework

1.1   A Capsule History

The first documented reference to a term suggestive of the presently conventional meaning of the word *engineer* or *engineering* can be traced to early Roman times when the Latin expression *ingenium* was used to suggest some ingenious attribute of an object or a person. Soon after, this term was specialized to characterize an ingenious device useful for exceptional and important purposes. Eventually, its derivative *ingeniator* was applied to a person possessing an innovative mind and skillful hands in the making of such devices.

Though the Romans may be credited with assigning a label to individuals who were both clever and dexterous in the making of useful devices, they did not invent the practice of making ingenious devices. The primal instinct for innovative artifacts and the skills required in their making, had emerged during the earliest stirrings of the human imagination. Indeed, one may identify a long and systemic progression of such activities as suggested by the following: from prehistoric making of stone tools to the building of ancient pyramids; from medieval construction of cathedrals and fortresses to the intricate crafting of mechanical clocks and moveable-type printing; from isolated discoveries of glass lenses and iron casting to the development of the tower mill and steam power; from automobile assembly plants to modern jet aircraft manufacture, and more recently to deep-space probes and microchips. Throughout this vast range of adaptive progressions, innovative thought and skilled actions in the making of ingenious devices were essential to the emergence of communities, cultures, and civilizations. Hence, one may well assert that by demonstrated practice rather then by specific label, *ingeniatore* have a long and stimulating history.

About 400 years ago, the designation *engineer* as a substitute for *ingeniator* gradually appeared in common use. Then, some 200 years ago, a movement arose to endow the theory and practice of engineering with a more visible form of organized professionalism. With time, this included an emphasis on engineering identity, sharing of technical knowledge, standardization of practice, systematized instructional curricula, and projection of a public service function. These and similar considerations continue to evolve even today.

Engineering is now a highly technical and continuously evolving academic and professional discipline taught at universities and polytechnic institutions. Its foundational knowledge rests on parts of the natural sciences, its methodological foundations is based on rational thought, its obligatory justification relates to a societal interest, and its specialized professional practice is circumscribed by accreditation and licensing provisions. A variety of industrial organizations have emerged simultaneously and synergistically with the evolution of engineering, thereby providing a vast range of products and services to individuals and institutions. Indeed, engineering may now claim to be both a shaper and a reflector of contemporary times.

1.2   Core of Engineering

The innovative Roman *ingeniatore* were evidently engaged not only in tactile actions but also in a cognitive process. They must have thought about properties of natural materials, considered devices for specific purposes, proceeded by trial and error, and learned from failures. Now and then a particularly useful and ingenious device emerged and was duly noted by some writers of the day. Thus, an *ingenium* came about because of the thought and actions of some innovative *ingeniator* who had become familiar with the properties of natural materials and then imaginatively shaped and assembled them in particular ways so as to yield a device of particular utility. An astute observer could conclude that this interactive endeavor constituted a technical process involving naturally available materials which an *ingeniator*, by thought and skill, shaped and combined to create an *ingenium* of subsequent interest. This dynamic could be modeled as an input-output process

$$materials \rightarrow \boxed{ingeniator} \rightarrow ingenium \qquad (1.1)$$

with the rectangle suggesting creative thought and skilled actions of the *ingeniator* in combining natural materials and phenomena into an ingenious and useful device. Note however that this symbolic depiction is evidently both primal and universal since even for Stone Age Man we may write

$$\begin{matrix} natural \\ stone \end{matrix} \rightarrow \boxed{\begin{matrix} human\ action \\ of\ stone\ chipping \end{matrix}} \rightarrow \begin{matrix} stone \\ tool \end{matrix}, \qquad (1.2a)$$

as well as for a contemporary engineer

$$\begin{matrix} natural \\ materials \end{matrix} \rightarrow \boxed{\begin{matrix} engineering\ thought\ and\ actions \\ of\ design\ and\ manufacture \end{matrix}} \rightarrow \begin{matrix} computer \\ chip \end{matrix}. \qquad (1.2b)$$

Hence, and with considerable generality, a progression of much engineering relevance, may be typically characterized as

$$\begin{matrix} natural\ materials \\ and\ their\ properties \end{matrix} \rightarrow \begin{matrix} creative\ thought \\ and\ skilled\ actions \end{matrix} \rightarrow \begin{matrix} ingenious\ and \\ useful\ device \end{matrix} \qquad (1.3a)$$

with the rectangular enclosure omitted. A label and component listing is suggested by

$$Nature \begin{Bmatrix} materials \\ phenomena \end{Bmatrix} \rightarrow Engineering \begin{Bmatrix} thought \\ actions \end{Bmatrix} \rightarrow Devices \begin{Bmatrix} ingenious \\ useful \end{Bmatrix}, \qquad (1.3b)$$

or, to explicitly recognize time as the pervasive variable and using a functional notation to be used hereafter, we write

$$N(t) \begin{Bmatrix} materials \\ phenomena \end{Bmatrix} \rightarrow E(t) \begin{Bmatrix} thought \\ actions \end{Bmatrix} \rightarrow D(t) \begin{Bmatrix} ingenious \\ useful \end{Bmatrix}. \qquad (1.3c)$$

Finally, we represent these various expressions in compact form as

$$N(t) \rightarrow E(t) \rightarrow D(t). \qquad (1.3d)$$

We take this three component dynamic relation as primal to engineering and comment on the meaning of these constituent terms, as perceived in contemporary times, in the following sections. But first a comment about notation.

1.3 Symbolic Notation and Engineering

The preceding has introduced some conceptual constructions and associated symbolic notation of particular relevance to *Engineering in Time*.
Such symbolic notation is useful in our context for three reasons:

(a) *Familiarity*
    It represents aspects of thought and actions in graphical form which
    are familiar to students of all engineering disciplines.
(b) *Empiricism*
    It constitutes a practical way of depicting complex processes of cen-
    tral importance to the theory and practice of engineering.
(c) *Heuristics*
    It suggests useful and efficient means of organizing and exploring
    the historical progression and contemporary context of engineering.

We emphasize that symbolic and graphical notation has long proven
to be a most effective and powerful instructional and operational tool in
engineering. While such notation can take many forms — for example
force diagrams, circuit representations, electromagnetic field depictions,
vector notation, phase-plane projections, etc. — its pedagogical power
rests in the effectiveness of graphical-geometrical depictions providing
a valuable cognitive focus for complex physical phenomena and processes. Indeed, engineers have become particularly adept at this type of
visual and imaginative thinking and it is for this reason that symbolic
and graphical notation is commonly introduced and will be also here be
used, adapted, and expanded[†].

1.4 Essential Components

The leading term $N(t)$ in Eq. (1.3d) represents nature as the basic starting
point of a most relevant progression. For purposes of elucidating this
progression, $N(t)$ may be characterized by several attributes:

(a) Nature constitutes the resource of basic materials essential to
    the practice of engineering (e.g. stone, wood, fiber, water, metal,
    petroleum, . . .).

[†]Appendix A provides a brief discussion of mathematical ideas and notation in history.

(b) Nature embodies the physical phenomena of foundational importance to the theory of engineering (e.g. hardness, diffusion, heat,
elasticity, friction, turbulence, . . .).
(c) Nature possesses autonomous dynamical features of relevance to the
performance of engineered devices (e.g. earthquakes, water cycle,
hurricane winds, flash floods, . . .).

In Sec. 9.7, we add two additional but highly personal and private features to this list: nature may induce a transcendental sense of *engagement*
and it may also stimulate a deep sense of *place-attachment*.

The term $D(t)$ in Eq. (1.3d) identifies engineered devices — the ingenious and useful human-made objects so judged by common pragmatic
criteria. Contemporary engineering practice suggests the following as
helpful working definitions for this term:

(a) A device may be an artifact, constructed or adapted but invariably
based on considerations of synthesis; it may be a hardware object
such as a tool, prosthesis, sensing instrument, interactive machine,
small-scale appliance, large-scale assemblage, adapter/converter,
passive/dynamic network, etc., or it may be a cognitive object such as
a strategic idea, optimal process, action program, information management schema, heuristic algorithm, software package, experimental know-how, etc.
(b) A device may be associated with a commodity providing selected
functions: projection of information (e.g. book, clock face, moni-
tor display, . . .), means of transportation (e.g. wagon, parachute, aircraft, . . .), serve the needs of sustenance (e.g. water pipe, cutlery,
flour mill, . . .), supply entertainment (e.g. guitar, video, dynamic
art, . . .), provide physical comfort (e.g. furnace, lawn chair, mosquito
spray, . . .), etc.
(c) A device may also be characterized as a synthesized object which
engages, *stimulates* and *enables* its makers and users.
(d) A device may further be viewed as a convenient unit of engineering
theory and practice.

These four characterizations evidently span considerable breadth
and hence provide much opportunity for association with specific
examples.

With $N(t)$ and $D(t)$ so characterized, it is evident that engineering
$E(t)$ constitutes the central and uniting connection in the progression
from $N(t)$ to $D(t)$, Eq. (1.3d). A definition and two corollaries for

engineering may now be introduced:

(a) *Definition*

Engineering represents creative thought and skilled actions asso-ciated with the use or adaptation of natural materials and natural phenomena in the conceptualizing, planning, designing, develop-ing, manufacture, testing, implementing, improving, and disposing of devices.

(b) *Corollary I*

Engineering may be interpreted as the rational activity concerning specifics of *how to make devices* and *how they might be made better*.

(c) *Corollary II*

Engineering constitutes the domain of thought and action which uses *what is* to create *what may be*.

Note that these characterizations evidently define engineering not only as a *creative* activity but also as a *goal-seeking* activity. Table 1.1 provides a summary characterization of the primal.

Table 1.1   Component characterization of the primal $N(t) \rightarrow E(t) \rightarrow D(t)$.

$N(t)$	$E(t)$	$D(t)$
1. Resource	1. Creative thought	1. Material artifact
2. Phenomena	2. Skilled actions	2. Cognitive object
3. Dynamics		3. Commodity support
4. Engagement		4. Maker/user stimulant
5. Attachment		5. Engineering unit

1.5   Change and Engineering

All contemporary professions need to consider change and all describe change in terms which are specific to their interests. Thus, economists speak of *business cycles*, chemists have found the notion of *reaction chains* useful, human resource managers consider *career progressions*, physicists express a number of dynamical processes using the concept of *phase transitions*, historians have identified *sequences of historical causation*, environmentalists have found *cascading sequences* as vital, biologists have for the past century devoted much effort to *biological evolution*, and mathematicians have introduced a specialty known as *stochastic methods*. Even popular terminology makes use of *chains of events* and the recognition that *the only constant in life is change*. Indeed, changes are foundational to life and living.

For engineers, the term *progression* is especially informing in the characterization of some important changes in their work. To begin, relations (1.1) to (1.3) suggests features well-known in engineering such as input → output processes involving materials, energy, ingenuity, and information. Our interest here, however, is more extensive and we view such relations and associated expressions as a catalyst to suggest other time-ordered connected terms central to our conceptualization of *Engineering in Time*. To clarify this emphasis we consider four classes of progressions.

1.5.1 *Homogeneous Progressions*

As a first illustration, consider an aircraft cleared for take-off. Evidently it may have its subsequent dynamical features specified by the homogeneous progression

$$\begin{matrix} zero\ speed \\ at\ time\ t_0 \end{matrix} \rightarrow \begin{matrix} intermediate\ speed \\ at\ time\ t_1 \end{matrix} \rightarrow \begin{matrix} take\text{-}off\ speed \\ at\ time\ t_2 \end{matrix} \rightarrow \cdots . \quad (1.4a)$$

As another illustration, the sequential changes of the mean temperature in a reaction vessel in the process of start-up or shut-down may be represented in general as

$$T(t_0) \rightarrow T(t_1) \rightarrow T(t_2) \rightarrow T(t_3) \rightarrow T(t_4) \rightarrow \cdots . \quad (1.4b)$$

These two examples represent homogeneous progressions and involve some specific measurable variable $X(t)$ at successive time coordinates; in complete generality we may write for such changes in time

$$X(t_0) \rightarrow X(t_1) \rightarrow X(t_2) \rightarrow X(t_3) \rightarrow \cdots . \quad (1.4c)$$

Thus, homogeneous progressions describe the magnitude of a time-dependent variable which is associated with some device or devices. Note that in a homogeneous progression each term possesses the same units.

1.5.2 *Heterogeneous Progressions*

The practice of engineering involves many informing progressions which
are not fully describable by one homogeneous variable, Eq. (1.4). Consider the following three historical examples:

$$\textit{chariots} \rightarrow \textit{wagons} \rightarrow \textit{trains} \rightarrow \textit{automobiles} \rightarrow \cdots \,, \qquad (1.5a)$$

$$\begin{array}{c}\textit{shadow}\\\textit{clocks}\end{array} \rightarrow \begin{array}{c}\textit{water}\\\textit{clocks}\end{array} \rightarrow \begin{array}{c}\textit{weight-driven}\\\textit{clocks}\end{array} \rightarrow \begin{array}{c}\textit{pendulum}\\\textit{clocks}\end{array} \rightarrow \begin{array}{c}\textit{spring}\\\textit{clocks}\end{array} \rightarrow \cdots \,,$$

$$\qquad (1.5b)$$

$$\begin{array}{cccc}\textit{animal} & \textit{water} & \textit{wind} & \textit{steam}\\\textit{power} \rightarrow & \textit{power} \rightarrow & \textit{power} \rightarrow & \textit{power} \rightarrow \cdots \,.\\(\sim 10^4 \; BP^\dagger) & (\sim 3000 \; BP) & (\sim 700 \; CE) & (\sim 1700)\end{array}$$

$$\qquad (1.5c)$$

In addition to these explicit retrospective expressions, some general
engineering projective progressions are suggested by the following:

$$\textit{opportunity} \rightarrow \textit{invention} \rightarrow \textit{enhancement} \rightarrow \cdots \,, \qquad (1.5d)$$

$$\begin{array}{c}\textit{innovative}\\\textit{proposal}\end{array} \rightarrow \begin{array}{c}\textit{design}\\\textit{study}\end{array} \rightarrow \begin{array}{c}\textit{prototype}\\\textit{testing}\end{array} \rightarrow \begin{array}{c}\textit{device}\\\textit{modification}\end{array} \rightarrow \cdots \,, \quad (1.5e)$$

$$\begin{array}{cccc}\textit{emerging} & \textit{responding} & \textit{engineering} & \textit{plans}\\\textit{societal} \rightarrow & \textit{government} \rightarrow & \textit{project} \rightarrow & \textit{and} \rightarrow \cdots \,.\\\textit{interest} & \textit{policies} & \textit{organization} & \textit{proposals}\end{array}$$

$$\qquad (1.5f)$$

In distinction to the general homogeneous progression (1.4c), the progressions (1.5a)–(1.5f) are evidently heterogeneous and may be written
in the general form of

$$A(t_0) \rightarrow B(t_1) \rightarrow C(t_2) \rightarrow D(t_3) \rightarrow \cdots \,. \qquad (1.5g)$$

While the meaning of the terms of homogeneous progressions (1.4c)
generally follow from some appropriate differential equation, no com-
parably compact and explicit defining expression can be specified for
the terms of the heterogeneous progressions (1.5g). What can however
be done is to enumerate some relevant features of such heterogeneous

---

†BP: Before Present in units of years, Appendix B.

progressions of specific interest to engineering and to our objectives in this text:

(a) The initial stimulant $A(t_0)$ generally determines the nature of the progression
(b) The terms may relate to a class of devices, a class of functions, or various combinations
(c) The terms may possess a range of sequential dependencies or strength of connections
(d) The terms may relate to specifics of experience and/or knowledge-base of participating individuals
(e) The terms may be influenced by the extent of cooperation or competition among various participating or affected individuals
(f) The terms may reflect upon selective preferences associated with affected societal institutions and motivating political ideologies.

Note also that what is actually *flowing* in a heterogeneous progression is a combination of forms of matter, energy, information, ingenuity, and authority.

Then, to further amplify the distinction between these two types of progressions, one may also conclude that homogeneous progressions generally tend to possess a predictive and inorganic character, while heterogeneous progressions possess a probabilistic and organic character; that is, heterogeneous progressions display the contingent role of the human imagination and community preferences.

Finally, observe that heterogeneous progressions are also used in other dynamical contexts; this includes business plans, strategic procedures, resource explorations, institutional development, optimality scheduling, and critical-path programming.

It is becoming increasingly essential that engineers think in terms of heterogeneous progressions for such evolutions invariably characterize important aspects of the practice of engineering.

1.5.3 *Primal Progression*

For our purposes here, we assign a very special meaning to the family of heterogeneous progressions represented by Eqs. (1.1)–(1.3). These three-node progressions with engineering explicitly indicated as $E(t)$ in Eqs. (1.3b)–(1.3d), or implicitly suggested as in Eqs. (1.1), (1.2), and (1.3a), will be labeled the engineering *primal progression* — or simply the primal:

$$N(t) \rightarrow E(t) \rightarrow D(t). \qquad (1.6)$$

We will repeatedly encounter this short progression as a focus for description.

1.5.4 *Connectivity Progression*

Further, there are good reasons to expand on the primal progression (1.6) by the addition of terms which suggest processes such as feedback, feed-forward, recursion, and branching; this expanded formulation will take on the appearance of a directed graph and, for purposes of terminological consistency, will be labeled the engineering *connectivity progression* — or simply connectivity:

$$\underbrace{N(t) \rightarrow E(t) \rightarrow D(t) \rightarrow (additional\ terms)}. \qquad (1.7)$$
$$(additional\ processes)$$

Note that while both Eqs. (1.6) and (1.7) are progressions, the primal (1.6) will serve as the enduring kernel of the connectivity (1.7). Table 1.2 provides a summary table of the four progressions of interest here.

Table 1.2   Summary tabulation of engineering progressions.

Label	Functional Form
Homogeneous	$X(t_0) \rightarrow X(t_1) \rightarrow X(t_2) \rightarrow X(t_3) \rightarrow \cdots$
Heterogeneous	$A(t_0) \rightarrow B(t_1) \rightarrow C(t_2) \rightarrow D(t_3) \rightarrow \cdots$
Primal	$N(t) \rightarrow E(t) \rightarrow D(t)$
Connectivity	$\underbrace{N(t) \rightarrow E(t) \rightarrow D(t) \rightarrow (additional\ terms)}$
	$(additional\ processes)$

We emphasize that engineering consists not only of a chronology of invention, innovations and specific engineering events — traditionally described in literary form — but also of a simultaneously evolving and increasingly complex pattern of the workings of engineers, well described by a connectivity based on considerations of heterogeneous progressions. A combined literary and symbolic connectivity description offers a more informing characterization of the state and evolution of engineering.

1.6 Engineering and Time

Engineering has, over time, become increasingly identified by association with classes of devices: Civil Engineering with *civic* devices, Mechanical Engineering with *mechanical* devices, Chemical Engineering with *chemical* processing devices, Electrical Engineering with *electrical–electronic* devices, and so on. The association between a specific engineering discipline and associated class of devices may be indicated by subscript notation to the primal progression,

$$N(t) \rightarrow E_i(t) \rightarrow D_{i,j}(t) \tag{1.8}$$

to suggest therefore the time-varying workings of an engineer associated with the ith engineering discipline in the making of ingenious j-type devices.

An evidently significant feature of the primal (1.8) is now apparent: devices $D_{i,j}(t)$ do not appear in isolation but emerge from a broad historical context and relate to conceivable engineering projections about the future. To highlight this creative and goal-seeking feature of engineering, we introduce the following explanatory interconnection:

$$N(t) \rightarrow E_i(t) \rightarrow D_{i,j}(t)$$

$$\updownarrow$$

$$\left\{ \begin{array}{l} \textit{Historical evolution} \\ \textit{of devices which relate} \\ \textit{to and predate } D_{i,j}(t) \end{array} \right\} \begin{array}{c} \searrow \\ \rightarrow D_{i,j}(t) \rightarrow \\ \nearrow \end{array} \begin{array}{c} \nearrow \\ \\ \searrow \end{array} \left\{ \begin{array}{l} \textit{Future development} \\ \textit{of devices which relate} \\ \textit{to and postdate } D_{i,j}(t) \end{array} \right\}.$$

$$\tag{1.9}$$

Thus, the primal progression embodies a unique evolutionary past and a contingent future, establishing thereby a distinctive temporal complexity to the evolution of engineering. Engineers need to understand the related past for their creative thoughts and skilled actions become operational in the future. An important component of professional engineering relates to the depth and breadth of understanding of this relation.

Evidently then, the range of heterogeneous progressions of interest here provides for a domain of preserved knowledge and experience so that

Engineering is Cumulative.

Then, the emergence and implementation of new devices is influenced by a variety of uncertain developments and therefore

Engineering is Contingent.

And finally, the simultaneous accommodation of selected domains of cumulative knowledge and experience in anticipation of a contingent future suggests a third feature, namely that

Engineering is Dynamic.

Engineering is thus a profession of considerable breadth and depth — and therein lies the challenge which has long attracted creative and skilled individuals.

1.7   To Think About

- Consider common devices such as bicycles, cars, household appliances, aircraft, etc. What events or developments have enhanced or could have frustrated their particular evolution?
- Progression (1.3d) appears like a conventional input → output model applicable to many areas of engineering. Explain the reasons for this similarity emphasizing in particular the thoughts and actions of engineers.
- Homogeneous progressions are generally describable using the calculus; in contrast, heterogeneous progressions are equally common but very difficult to describe analytically. Why? In addition to particular device developments, consider also specific historical subjects such as biography, corporate histories, national evolutions, etc.

PART B

History of Engineering

An interest in making ingenious devices is a primal human instinct dating to more than a million years in the past. With time, this developed into a powerful quest to use the creative capacity of the mind and manipulative skills of the hand to establish not only a chronology of engineering invention and innovation but also an evolving connectivity of defining relevance to the theory and practice of engineering.

CHAPTER 2

Prehistoric Engineering

$(\sim 10^6 \text{ BP} \rightarrow \sim 10^4 \text{ BP})$

Primal Discovery of Devices

2.1 Early Humans

The reconstruction of early human history has long been pursued by archeologists and anthropologists. Their work involves painstaking examination of a variety of earth's strata in search of rare skeletal evidence and surviving artifacts. Detailed study of such objects has yielded useful information on the characterization and practices of hominids for the past million years. Of particular interest is the available evidence suggestive of prehistoric makings of useful objects by humans, for therein one can recognize the primal stirrings of engineering, that is Prehistoric Engineering.

To begin, important early skeletal evidence of hominids (bipedal, upright) has been dated to a time between 3 and 2 million years ago. A widely publicized find corresponding to this prehistoric time occurred about 30 years ago in equatorial east Africa. The subject, named *Lucy*, possessed a brain volume of $\sim 600 \text{ cm}^3$ which may be compared to $\sim 1600 \text{ cm}^3$ for modern humans, Fig. 2.1.

Widespread prehistoric tool-making became especially evident with skeletal finds in the 1930s near Beijing, China. This so-called Peking Man, estimated to have lived about 500,000 BP, was short with a sloping forehead and a brain volume of about 1000 cm^3. Chipped stones and ashes from that site suggest that both stone tools and fire making were known to this early cave dweller.

A prior find — occurring about 1860 in the Neander Valley of Germany and subsequently confirmed elsewhere in Europe, Middle East, and North Africa — has identified Neanderthal Man. This human species existed between about 300,000 BP and 30,000 BP, was stocky, also with a sloping

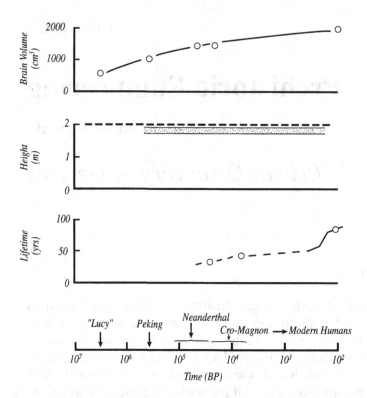

Fig. 2.1   Summary characterization of some average features associated with early and contemporary humans. Note the substantial increase in brain volume with time and that average height appears to be largely time invariant. Increased mean lifetime has been significant in recent times for reason of changes in sanitary installations and improvements in health care.

forehead and possessed a brain volume of ~1400 cm³. By archeological evidence, stone-tipped spears, wooden clubs, and variously chipped stones had by then been crafted. Thus, human-made artifacts had now become more diverse and specialized.

The next sequential prehistoric human to be noted is Cro-Magnon Man. Its skeletons were first found in the mid 1800s in the Cro-Magnon hills of southwestern France and subsequently throughout the World. Dated to an initial appearance of about 30,000 BP, contemporary human features of vertical forehead and brain volume had by then evolved. Artistic skills had also emerged as evidenced by the still observable wall paintings in ancient caves. Cro-Magnon people are generally viewed as our oldest anatomically similar ancestors.

2.2 Invention of Tools

Anthropologists point to persuasive evidence suggestive of some fundamental and universal human needs which are quite independent of any
state of cultural achievement; among the most basic such needs are those
associated with nutrition, security, hygiene, and social interaction including trade. Evidently, even the most rudimentary forms of human-made
devices relate to these fundamental needs so that, as a basic premise of
interest here, the heterogeneous progression

$$materials \rightarrow inventiveness \rightarrow devices \qquad (2.1)$$

constitutes a critical prehistoric human development: humans are profoundly endowed with the instinct to examine, to invent, to make, and to
use devices.

One may well imagine that somewhere in the mist of prehistory and
likely first in the east-central part of Africa, the earliest hominids began
using twigs to augment the actions of fingertips in their search for edible roots. Some time later, they must have discovered that this search
could be rendered more effective with particular twigs and that it could
be made even more efficient by chewing the end of twigs to produce a
suitable point. Then, by an enormous leap of the primitive imagination —
estimated to have occurred over a million years ago — it must have been
repeatedly discovered that it was easier and more efficient to produce
the desired pointed twig with some carefully directed blows of naturally
available sharp-edged stones. While these sharp stones might be found
by chance, primitive humans also discovered that suitably angled blows
of a hard stone (e.g. granite) onto a softer stone such as flint (i.e. fine-
grained siliceous grey rock) or obsidian (also known as black glass),
could produce a concave surface with very sharp edges. This extraordinary acquisition of a manual skill combined with a planned application,
created a device of considerable utility and durability thereby also identifying the emergence of the Stone Age. The primal progression (2.1) may
therefore be rendered more specific as a variation on Eq. (1.3):

$$N(t) \{wood\} \rightarrow E(t) \{chewed\ point\} \rightarrow D(t) \{wood\ tool\}, \qquad (2.2a)$$

and

$$N(t) \{natural\ stone\} \rightarrow E(t) \{slanted\ blow\} \rightarrow D(t) \{stone\ tool\}. \qquad (2.2b)$$

Thus, the earliest human actions in the making of devices involved hand-held objects forcefully directed onto a stone surface with material percussion yielding the desired results. The property of nature thus exploited for human purposes was the hardness and crystal structure of some materials.

The earliest stone tools were chopping tools, typically consisting of a grapefruit-sized smooth stone with flakes knocked off using a harder stone or even hardwood, Fig. 2.2. With time, these stone tools were selectively chipped to be useful for a variety of purposes such as shaping wood, cutting carcasses, cleaning hides, and boring holes in hide or wood. Thus, by the application of primitive engineering skills, early humans could expand their activities as an aid to survival and begin to explore and control their world.

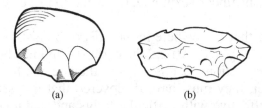

(a)   (b)

Fig. 2.2   Examples of earliest ingenious devices: (a) hand-held stone chopping tool, and (b) cutting and scraping stone tool. The act of making such prehistoric tools is now known as *stone knapping*.

A closer examination of the evolution of the works of these prehistoric stone-age tool makers also suggests a primitive instance of engineering synthesis: a sharp-edged elongated flaked stone tool could be tied with animal sinew to a sturdy wooden pole thereby forming a lance or a spear. Similarly, a stone chopping tool could be fastened to a short wooden club thereby forming an axe or hammer. Prehistoric humans thus discovered that something new and ingenious could be made by a suitable combination of something old according to the primal progression

$$N(t) \begin{Bmatrix} wood \\ stone \\ sinew \end{Bmatrix} \rightarrow E(t) \begin{Bmatrix} selecting \\ fitting \\ binding \end{Bmatrix} \rightarrow D(t) \begin{Bmatrix} spear \\ lance \\ hammer \end{Bmatrix} . \quad (2.3)$$

Note that this primal now appears like a vector or set progression, known in mathematics as operational mapping.

It is for reasons of such evolving specialization of stone tool development that the Stone Age has been further subdivided into the Paleolithic (Old Stone Age), the Mesolithic (Middle Stone Age), and the Neolithic (New Stone Age).

Sharp Stone Edges

A visual examination of prehistoric stone tools generally display many adjacent concave surfaces. Modern science attributes these so-called conchoidal surfaces to the ultrafine crystal structure of siliceous and obsidian materials. In these solids, a concentrated blow causes the formation of a characteristic fracture cone emanating from the point of impact. Hence, contemporary artisans are able to reproduce prehistoric arrow heads and spear points since the same kinds of materials still exist naturally. Also, it is because of these conchoidal surfaces that shattered glass and broken porcelain constitutes a common hazard.

Prehistoric people also learned how to keep animal hide from perishing. They used sharp stone scrapers to remove the decomposable material from the inside. Then, leaving the hide for sun drying would produce stiff material suitable for walls and roofs; however, these early crafters also discovered that soaking fresh hide in various solutions with subsequent kneading or chewing and treatment with fats could make hide sufficiently soft for clothing.

2.3   Discovery of Fire

The discovery of a means to produce and transport fire at will is widely viewed as a most epochal achievement of Stone Age humans. This use of a natural phenomena has proven to be of profound significance in the development of human adaptation and cultural evolution.

Peking Man may well have been among the first to establish the practice of using fire, for in the immediate vicinity where the skeletal remains were found there exist campfire ashes. Elsewhere in Europe, structures which appear to be prehistoric hearths have been discovered and dated to about the same time.

Actual steps taken in the attainment of the control of fire is evidently beyond research but once fire has been naturally started — by lightning or lava flow — it could have been sustained and transported by burning wood sticks and preservation of embers. Fire may have also been started by friction on hardwood to produce heat or by sparking between hard stone and flint, and the associated use of kindling such as wood chips and dry grass.

Control of fire affected early humans in most profound ways. Among those to be noted in particular are the following:

(a) *Protection*
All animals fiercely seek to avoid fire so that burning sticks and campfires can fend off predator attacks.
(b) *Nutrition*
Many foods are difficult to digest raw (e.g. raw vegetables, meats, ...) but may be rendered edible upon heating; heat also destroys parasites and bacteria.
(c) *Range*
The transport of fire enabled Stone Age people to migrate from tropical Africa to colder climates of northern Europe and Asia.
(d) *Community*
One may well image that evening campfires provided a suitable social environment for re-enactments of hunting experiences and for story telling. Hence, control of fire may have directly contributed to the formation of human expression and the establishment of mythical tales and legends.

In this case of fire, the property of nature found most beneficial to human life was the self-sustaining exoergic molecular rearrangement of some materials.

Risk of Fire

Prohibitions against arson and commonly experienced bodily pain from fire have established universal apprehensions about this natural phenomenon, even while its critical role in the natural cycle of wilderness plant sustainment is becoming increasingly recognized. How fire first became a transportable commodity continues to be unresolved though anthropologists now tend to view this development as both cognitive and tactile; only humans have acquired the capacity to control fire.

2.4 Customs and Art

There exists good evidence to indicate that the Neanderthals were among the first to develop some social customs and community practices, suggesting also the early emergence of tribal identity. Three particular practices seem to have emerged in these prehistoric times:

(a) *Burial*
   Some graves contain human skeletons aligned in a sunrise/sunset direction, implying therefore some specific beliefs associated with natural phenomena.
(b) *Hunting*
   Stone pits have been found containing mostly bear skulls, suggesting the first practice of collecting trophies associated with successful hunts.
(c) *Stampeding*
   Identification of piles of ibex bones at the base of cliffs and bottoms of wide crevices indicates the prehistoric practice of stampeding animals for food supplies.

In contrast to the roving Neanderthals, the subsequent Cro-Magnons were more settled cave and cliff dwellers and herein lies a notable artistic development. There exist many caves in France and Spain containing stone etchings and charcoal finger-paintings of animals with those deep inside still remarkably well preserved even though ~25,000 years may have elapsed. Nearly all such paintings contain considerable anatomical detail and informative shadings. Somewhat unexpected and puzzling, is the feature that many of these paintings are on walls and ceilings of narrow passages rather than in areas with more viewing space.

Thus the Prehistoric era is associated not only with the invention of primary tools and discovery of the control of fire but is also associated with the initial stirrings of social customs and artistic expression.

2.5 Shelter and Migration

Humans of the Prehistoric era were initially foragers and subsequently hunters. For reasons of survival they followed seasonal plant growth patterns and migrating herds of animals. In tropical zones, shelter was invariably provided by caves and overhanging cliffs and such locations have been productive sites for anthropological and archeological investigations.

Migration of early humans from their apparent primary-source region of equatorial east Africa proceeded to the Middle East and from there to Europe and Asia, Fig. 2.3. Archeological evidence indicates that these migrations into temperate zones were somewhat continuous, requiring adaptation with respect to clothing and shelter. Sites dated to about 100,000 BP suggest the use of large mammoth tusks arranged to form dome-like structures presumably covered with tree branches. Smaller shelters made of bones and hide have been dated to about 30,000 BP. Primitive building construction — like the making of stone tools — thus also reflects on early man's instinct to make useful things.

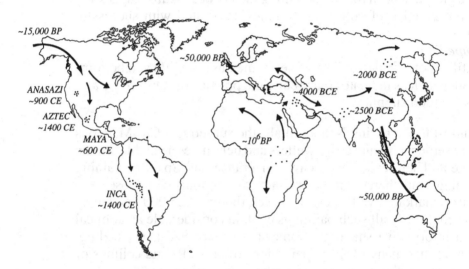

Fig. 2.3   Prehistoric making of devices spread by global migrations (→). Approximate dates for early population centers are indicated (⁂).

During the last ice age, reduced oceanic distances opened land bridges and island-hopping pathways for migration from Asia to Australia as well as from Europe to England, both about 50,000 BP. Similarly, from about 30,000 BP to about 15,000 BP, various bands of hunters followed game and crossed from Asia to North America along the land bridge in the Bering Strait region thus establishing — or contributing to — human occupation of North and South America. Subsequent melting of the polar ice caps raised the sea level by about 100 m thereby tending to isolate these nomads on continents and island of various sizes.

2.6 Prehistory and Invention

Stone tools of the Prehistoric era appear to have many independent inventors because similar tools have been found in various distant regions. There is also good reason to believe that — unlike present practice — Stone Age people made tools for personal use only as the need arose and these were subsequently discarded; hence the reason for many stone chipped tools found at habitation sites. With stones generally plentiful and their production simple, there was no need to be burdened by always carrying them around.

Furthermore, among the more than 10^6 stone tools from the Prehistoric era are on deposit in museums, there exist some rare and incomplete tools made from vegetable and animal matter. These suggest the bow made from wood or bone and strung with animal sinew, as well as the boomerang and the sling. Additionally, variously shaped hooks and harpoon points made from bone were also fashioned for various purposes. Prehistoric engineers were indeed adept in the application of available natural materials for useful purposes.

With prehistoric tools found throughout most of the world, it is evident that inventing is an instinctive trait of humans; details of invented devices and the extent of utilization however varies considerably with time and region.

2.7 Prehistoric Engineering: Discovery of Devices

This chapter has provided a description of the emergence of the earliest form of engineering, the Prehistoric period $\sim 10^6$ BP $\rightarrow \sim 10^4$ BP.

An important heterogeneous progression, now representing primal connectivities (2.2) and (2.3), has here been identified for Prehistoric times and needs to be stressed. With stone available in nature $N(t)$ by simple collection and then by process of percussion, the engineers of the day $E(t)$ created ingenious devices $D(t)$ for their own personal use as the need arose. Prehistoric Engineering may hence be well characterized by the engineering primal progression

$$N(t) \rightarrow E(t) \rightarrow D(t), \tag{2.4}$$

thereby representing the profound discovery of the making of devices.

Prehistoric Engineers are evidently the primal makers of ingenious devices and Eq. (2.4) provides a compact engineering connectivity characterization of the $\sim 10^6$ BP $\rightarrow \sim 10^4$ BP period, Table 2.1.

Table 2.1 Time axis illustration of Prehistoric Engineering ($\sim 10^6$ BP $\rightarrow$ $\sim 10^4$ BP). The several characterizations here introduced for this period of time in the Content Summary take on the form of a column vector; note, however, that the components are not independent.

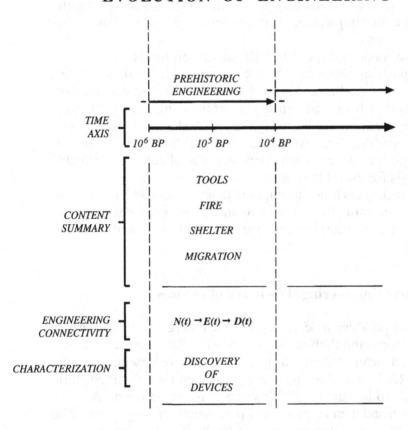

EVOLUTION OF ENGINEERING

2.8 To Think About

- Some ancient artifacts survive with time better than others. Why is this and what are the important conditions and underlying physical processes?
- Fire is a means of generating thermal energy. Why does this natural phenomena render some food edible and also destroy some bacteria?
- Itemize uses and applications possible with various designs of simple stone tools.

CHAPTER 3

Ancient Engineering

(~8000 BCE → ~500 CE)

Societal Interest in Devices

3.1 Human Practices

By about 20,000 BP, the global population consisted of an estimated two million foragers and hunters, leading a generally haphazard nomadic existence and possessing only objects which could be readily carried. But then small settlements and population clusters began to emerge in verdant canyons and alluvial flood plains evidently aided by the appearance of wild but edible cereals. The Middle East and more specifically the area between the Tigris River and Euphrates River of present-day Iraq — often referred to as Mesopotamia which means *The Land Between Two Rivers*, Fig. 3.1 — provided climatic and soil fertility conditions particularly suited to stimulate changes in human practices. About 12,000 BCE, the dog began associating with humans while sheep, goats, and cattle became domesticated beginning about 10,000 BCE[†]. Wild barley then became a food staple and the initial stages of plant domestication were then undertaken. And so, the first of many riverine civilizations began to emerge.

It is estimated that by about 8000 BCE, some 500,000 people inhabited Mesopotamia. With increasing plant and animal domestication now extended to wheat and cattle, humans tended to become herders, then settlers, and finally tillers of the soil. The heterogeneous progression

$$foraging \rightarrow hunting \rightarrow herding \rightarrow settling \rightarrow soil\ tilling \qquad (3.1)$$

[†]The horse appears to have been domesticated in the steppe regions of Asia and Southeastern Europe, about 5000 BCE.

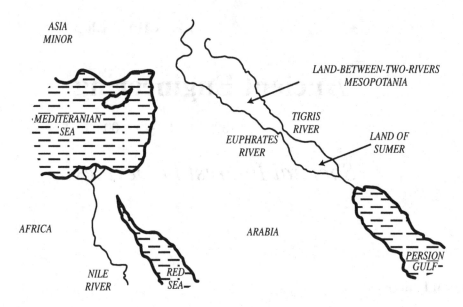

Fig. 3.1   Regions of relevance to the beginnings of Ancient Engineering.

identifies the principal human activities and the emergence of agriculture; one might claim that the discipline of primitive Agricultural Engineering was thus born. This progression also involved a substantially decreasing land-area-per-person needed to meet nutritional requirements and — most significantly — their invented devices now became not easily transportable and hence more plentiful. The term Agricultural Revolution is often used as a label for this major change in human practices.

Of Spinning and Weaving

Lost somewhere in the fog ancient Middle East history are the vital beginnings of spinning and weaving. It must have involved fleece of fluffy sheep wool from which continuous drafts were pulled and twisted with the aid of a freely rotating whorl. By tying and interconnecting, a crude weave emerged. Spinning skills were subsequently extended to the use of hemp and flax for the making of rope. Primitive looms may have thereupon been made to aid in making blankets, clothing, and eventually sails.

Changes from a nomadic gatherer-hunter lifestyle to one of settled herding and soil tilling involved the development of a number of innovative devices and practices. Not only were techniques of tool making extended, but also skills of plant selection, seeding, irrigating, and harvesting became increasingly developed. To be noted is the scratch plow — a vee-shaped tree branch for pushing and pulling — to loosen the top 3–5 cm of soil for planting. Further, artifacts resembling sickles consisting of sharp-edged flint inserted in slots of curved wood have been found as evidence of increasingly efficient tool development for harvesting.

During this Agricultural Revolution, the first thermally-induced material transformation was discovered. One may suggest that coarse weaving using vines and water-soaked twigs were covered with clay. Upon sun drying, these mats and containers yielded rigid shapes providing containers for holding small kernels and even liquids. The need for weaving was eventually eliminated by about 5000 BCE when it was found that campfire heat would irreversibly harden the clay to yield crude pottery.

It was also discovered that if straw is mixed with moist clay and the resultant blob molded into a more regular shape and allowed to sun-dry, an adobe brick useful for wall construction resulted. And thus, the acquired human skills as the facilitating mechanism provided for the important primal progression of Ancient times

$$N(t) \begin{Bmatrix} clay, fiber, \\ solar, heat \end{Bmatrix} \rightarrow E(t) \begin{Bmatrix} mixing, shaping, \\ fire\ heating \end{Bmatrix} \rightarrow D(t) \begin{Bmatrix} pottery \\ brick \end{Bmatrix}$$

(3.2)

yielding devices of increasing utility. This use of clay + straw + water adobe brick making, beginning about 5000 BCE, was soon followed by mortar consisting of sand and powdered limestone which, when mixed with water, served both as wall plaster and brick bonding agent[†].

But already during this earthenware developmental stage, the first signs of a metallicware development were stirring in Mesopotamia. Early humans must have long observed small shiny beads or chips occasionally found in creek and river beds, now recognized as naturally occurring alluvial nuggets of gold, silver, and copper. Because these objects appeared shiny, were malleable, and proved to be inert to decomposition, they were soon collected and shaped to be used for personal embellishment and community rituals.

[†]Early crafters had also used tar/bitumus materials mixed with sand as mortar.

Further — one may imagine — when a charcoal fuelled fire happened to be located on some exposed copper ore or nuggets, a red-brown liquid flowed and when cooled formed little light-brown globules. Thus, by accident and probably occurring repeatedly in different areas, smelting was discovered thereby establishing the specific primal progression

$$N(t)\{Cu\text{-}ores\} \rightarrow E(t)\{smelting, \, shaping\} \rightarrow D(t)\{copper \, ornaments\}. \tag{3.3}$$

Development, however, did not end with copper smelting. When this copper-bearing ore was accidentally mixed with a black tin-bearing ore in a draft-blown fire, a hard and chip-resistant new metal resulted providing another variation on the primal

$$N(t)\{Cu, \, Sn\} \rightarrow E(t)\{draft \, heating, \, shaping\} \rightarrow D(t)\{bronze \, tools\}. \tag{3.4}$$

This new metal could be cast and held a sharp edge very well providing therefore further material choices in the making of devices; it also introduced a new age, the Bronze Age, beginning about 3500 BCE and thus superceding the Stone Age.

3.2   The Sumerians

Beginning about 4500 BCE, a most significant series of developments occurred when a group of people called the Sumerians began settling in the lower reaches of the Tigris River and Euphrates River, Fig. 3.1. These people possessed a special aptitude and interest in the continued making of ingenious devices and, in particular, they improved on pottery and weaving and also developed kiln-fired pots and bricks. They were the first sufficiently organized people to begin erecting large ceremonial structures and improve agrarian productivity by the construction of stone and brick lined irrigation canals.

About 3500 BCE, settlements became larger leading also to the first evidence of city-states, large terraced buildings, and the appearance of rulers as priest-kings. A sense of adoration and submission emerged focussing not only upon their community practices and values but also upon supernatural forces which, they must have believed, had an effect on harvests and hence on a more secure existence.

Also, about 3500 BCE, the Sumerians are credited with two most profound inventions. One was the wheel and axle, with the wheel initially constructed of solid wood, Fig. 3.2; its use for two-wheel carts and — with tamed horses by then available — chariots soon followed. The other

Fig. 3.2   Examples of the two most profound inventions of the Sumerians: the wheel and Cuneiform script.

innovation was a form of writing now known as Cuneiform script. This script involved obliquely cut reed pressed onto soft clay tablets which were subsequently hardened by heating, Fig. 3.2. Special symbols were adopted for particular items such as grain or goats to identify ownership and other symbols were used to identify 1, 2, 6 and 60 as particular numbers of objects; it is also surmised that eventually symbols became associated with specific syllabic sounds thus alluding to the idea of a phonetic alphabet. Scribes emerged and acted as the efficient record-keepers and accountants in trade and public records.

The Wheel

As would be expected, there exists total uncertainty about the circumstances leading to the invention of the wheel, about 5500 years ago. It could conceivably have been inspired by the rotating motion of grinding grain with chunks of flat sandstone, subsequently rounded and adapted as a crude pottery wheel. It has also been suggested that the wheel and axle became developed only after tame horses had become available, thereby providing for the emergence of the chariot.

These two most primal developments may be represented as

$$N(t)\{wood\} \rightarrow E(t)\{shaping, assembling\} \rightarrow D(t)\{wheel\}, \quad (3.5a)$$

and

$$N(t)\{clay, reeds\} \rightarrow E(t)\{standardization, notation\} \rightarrow D(t)\{clay\ tablets\},$$
$$(3.5b)$$

establishing thereby a basic form of both Transportation Engineering and Information Technology — about 5500 years ago.

With numbers and counting becoming recognized concepts, the Sumerians also began measuring time with the lunation of $\sim$29 days as a convenient reference time-interval. Table 3.1 provides a summary listing of the above and other achievements of this remarkable civilization.

One may also suggest that the Sumerians were the first people who by their skills of irrigation and adobe construction, as well as in casting and molding of bronze products, had discovered primal forms of Civil Engineering and Metallurgical Engineering.

Table 3.1   Listing and characterization of contributing achievements of the Sumerians, $\sim$4500 BCE $\rightarrow$ $\sim$3000 BCE.

Agriculture:	irrigation, planting, harvesting, ...
Domestics:	spinning, weaving, improvements in pottery, ...
Communication:	cuneiform script, clay records, accounting, ...
Metals:	copper ornaments, bronze tools, ...
Structures:	wheels and carts, rafts, large ceremonial buildings, ...
Time:	cyclic seasons and lunations, ...
Organization:	Priests/Kings, religious rituals, hierarchical organizations, ...

3.3   Settlements and Civilizations

Beginning about $\sim$3500 BCE the Sumerians can be considered to have advanced beyond basic subsistence thereby enabling the identification of an early civilization: cities, written language, organized hierarchical governments, unique community practices, and shared beliefs. Interestingly, population clusters leading to influential centers of civilization also emerged elsewhere during this Ancient era.

About 3500 BCE, a distinct group of people became unified along the banks of the Nile River, eventually becoming known as the Egyptian Civilization. Egypt is often called the Gift of the Nile and this for good reason. This long and stately river possesses remarkable regularity with an annual silt-laden flood providing natural large-scale irrigation and natural fertilization which, together with a sub-tropical climate, provided for several crops a year. With such an effortless supply of food available, Egypt came to possess a substantial excess of human labor which its rulers used for the construction of increasingly elaborate temples, laboriously hewn into rock and with large wall surfaces and tall stone columns expertly decorated with hieroglyphics. However, the Pharaoh's grand building ambitions were most conspicuously met by royal mausoleums in the form of pyramids. The technological requirements for these sandstone and granite structures also involved specialized tools and construction techniques such as bronze chisels, bow drills, bronze saws, surveying equipment, and effective means for large stone transportation — using rope, sleds, crude cranes, and occasionally rollers but no wheels, no pulleys, and no draft animals.

The Nile also proved important in other ways. Being wide and placid it served well in the development of boats and naval skills. Notable in this marine development was the square and triangular (lateen) sail and various types of riggings. Steering involved two side oars symmetrically placed aft.

Additionally, the regularity of the seasonal floods led the early Egyptians to obtain the first remarkably accurate estimate of the length of the solar year: peak floods were determined to occur on average about 365 days apart. Further, the shadows from their large obelisks also led the Egyptians to use *shadow sticks* as crude sundials with daylight time variously divided into 8 or 12 equal intervals.

Calendrics

Harmonizing the ancient Egyptian Solar Calendar — 12 periods of 30 days each with 5 added mythical days — with the Sumerian ~29 day lunar month proved to be troublesome and even today some communities use both measures of time, the former for secular purposes and the latter for some religious purposes. The reason for this problematic is numerical: one number is not a simple multiple of the other; presently accepted values for these two time intervals are ~365.2422... days and ~29.5306... days, respectively. The subsequent 7-day week cyclic interval apparently resulted from seven stellar objects of particular interest to ancient people of the Middle East.

Another large population cluster began about 2500 BCE in the Indus River Valley of present day India and Pakistan. These urban centres were laid out in a grid pattern with a centrally located citadel. Surrounding their cities were farming areas and, like the Sumerians, they also practised irrigation to support agriculture. But unlike Sumer and Egypt which soon established trade connection, the Indus River civilization — now generally associated with the culture of Ancient India — remained largely isolated.

Finally, another population cluster began about 2000 BCE, in the Yellow River Valley of present day China. Also isolated and displaying a similar urban-rural pattern to those along the Indus River, this population cluster eventually proved to be particularly influential for its early and significant discoveries: ink, $\sim$2000 BCE; lodestone magnet, $\sim$1000 BCE; paper, $\sim$200 BCE; porcelain, $\sim$500 CE; and gunpowder, $\sim$600 CE.

Often overlooked is the use of animal power for human purposes, beginning in Ancient civilizations. The anatomy of cattle and water buffalos is such that a head or shoulder yoke allowed the pulling of loads on skids or wagons. Animal power thus multiplied human power in the continuing development of agriculture. Indeed, the increasing use of non-human power represents a kind of *driving force* in human affairs.

A heterogeneous progression which describes these various settlement developments is suggested by

$$
nomads \rightarrow \begin{matrix} settled \\ agriculturalists \end{matrix} \rightarrow \begin{matrix} animal\text{-}powered \\ agriculture \end{matrix} \rightarrow \begin{pmatrix} Civilizations: \\ customs \\ language \\ cities \\ \vdots \end{pmatrix}.
$$

$$(3.6)$$

Figure 2.3 of the preceding chapter provides a global geographical perspective of the location of these population centers.

3.4   Structures and Symbols

The people of Sumer and of Egypt produced some of the earliest large engineering construction projects. First were irrigation channels initially consisting of simple criss-crossing furrows for flowing water, some of which were eventually made larger and deeper to provide limited storage and supply smaller canals. By about 3500 BCE, irrigation canals in Sumer exceeded 50 km in length and stone dams were constructed to impound flood water for subsequent irrigation. In Egypt, variations in land elevation were used to direct part of the Nile water over large areas and a number of clever devices were introduced for raising water; this included the remarkably effective balanced-beam *shadoof* and a range of human or animal powered vertical wheels with tilting water containers mounted along the rim.

While large irrigation projects met the agricultural requirements of its associated community, a societal interest also arose for large structures to serve a symbolic purpose — religious and political. One of the earliest known structures were the Sumerian ziggurats in the lower Tigris-Euphrates watersheds, consisting of successively terraced and recessed layers of stone and adobe brick, and a shrine at the top. The most notable was the Ziggurat at Ur reaching a height of 26 m and a base of 60 m × 70 m, Fig 3.3.

In Egypt, pyramidic symbolic structures began as oversized brick and stone walled tombs becoming successively larger and eventually forming pyramids as large funerary monuments. Some 80 large pyramids had been erected in near proximity of the Nile. These structures reached their apex about 2500 BCE with the three Great Pyramids at Giza near Cairo, Fig. 3.3, a legacy to the three Pharaohs of the Fourth Dynasty. The larger pyramid has a 230 m × 230 m square base and rises nearly 150 m; it possesses remarkable alignment precision suggesting exceptional surveying skills by these ancient Egyptian engineers. An average of about 25,000 workers are estimated to have labored for about 30 years in order to create each of the monumental royal tombs.

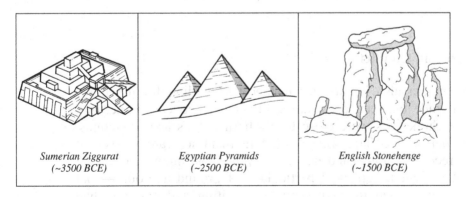

| Sumerian Ziggurat (~3500 BCE) | Egyptian Pyramids (~2500 BCE) | English Stonehenge (~1500 BCE) |

Fig. 3.3   Grand structures of Ancient Civilizations (not to scale).

Ancient Engineers

Ancient records and artifacts have been found which identify some
of the earliest engineers. Dating to the time of about 2900 BCE, a
man known as Imhotep is listed as the builder of some early pyramids in Egypt; he was also a government minister, a priest, and
physician to the Pharaoh of the day. Then, excavations in Sumer
have revealed a small clay statue of a man sitting with a scaled
plan of a temple and some drafting instruments on his lap; dated to
about 2300 BCE, this statue identifies the builder-engineer Gudea
who was also a priest-governor and who was deified after death —
presumably because of his piety and wisdom (and engineering
skills?).

The next exceptional community building project occurred in Europe
now commonly known as Stonehenge, Fig. 3.3. Located some 100 km
west of London, archaeological investigations suggest a construction
period of about 1000 years beginning about 2000 BCE. The geometry is concentric with originally 80 large vertical stones weighing
up to 50 metric tons and supporting horizontal lintels each weighing about 4 metric tons. All stones were quarried some 300 km north
west and laboriously transported by land and water and finally carefully erected. The reason for Stonehenge are hotly debated and vary
from a memorial to a Druid battle to Ancient Celt astronomical
markers.

3.5   Trade and Conflict

This Ancient era also led to circumstances providing for the introduction of trade between civilizations. It began between Sumer and
Egypt. The Sumerians were close to northern forests and mountains,
cultivated barley and flax, were skilled in metallic ware, possessed
means of record keeping, and generally were most innovative. Egypt
had wheat — which kept well particularly if ground as flour — but
little wood, possessed much papyrus for writing and painting, had
acquired skills for making bronze stone cutting and surveying tools,

and had developed techniques for sail and boat construction. Interestingly, rather than engaging in trade directly, the useful services of intermediate agents emerged. Present-day Lebanon was then populated by a group of people called the Phoenicians who adapted Egyptian merchant ships for Mediterranean travel thereby also becoming skilled sailors as well as expert traders; eventually they extended this function to other cities around the Mediterranean shores. Gold now also became the common currency in trade; this precious metal — it is valuable because it is rare and does not degrade with time — was panned in the head waters of the Taurus Mountains and mined in southern Egypt.

Sailing the Mediterranean was first routinely accomplished by the Phoenicians who had learned to chart their travel at night by the stars. By their various contacts, they also served as conduits of speech, language, and writing. Their exceptional contribution was the introduction of symbols for 22 consonants, written horizontally right-to-left. This provided humankind with the enduring concept of a phonetic alphabet, about 3000 years ago, to which the Greeks subsequently added vowels which eventually served as the basis of European alphabets.

While the establishment of techniques for the development of devices served well the interests of trade, such activities also proved decisive in warfare. For example, about 2000 BCE, Sumer was invaded by a people called Akkadians who had became expert makers and users of devices such as the bow and arrow; the Sumerians, however, had only developed the lance as a weapon and hence were easily conquered. Similarly, about 1700 BCE, the Egyptians were defeated by an invading people named Hyksos who had become very proficient in the use of two-person chariots, while the Egyptian weaponry consisted primarily of spears and clubs; interestingly, the suppressed Egyptians eventually also learned how to use chariots and 150 years later forced the Hyksos invaders out.

The discussion of this section suggests the primal progression

$$N(t)\left\{\begin{array}{l} stone, metal, \\ wood, clay, fire \end{array}\right\} \rightarrow E(t)\left\{\begin{array}{l} insight, skill, \\ initiative \end{array}\right\} \rightarrow D(t)\left\{\begin{array}{l} structures, tools, weapons, \\ tradeable\ goods, \ldots \end{array}\right\}$$

$$(3.7)$$

as a compact description applicable to Ancient times.

Ancient Government Practices

Warfare of Ancient times invariably involved plunder and slavery
of the defeated so that national wealth and the pool of skilled labor
were greatly affected by military conflict. In peace-time, a range of
regulations were sometimes introduced governing various societal
practices. Among the first written edicts containing articles specifically directed to engineers, are attributable to the Babylonian King
Hammurabi, issued about 1750 BCE. One section specifies that
if building collapse caused bodily harm to the owner then a proportional punishment — lashings, amputation, or enslavement —
was to be inflicted upon the builder; the practice of engineering in
Hammurabi's kingdom was evidently not for the timid.

3.6   From Minoan to Roman Times

By about 3000 BCE, the riverine civilizations of the Euphrates–Tigris
and the Nile had become important population centers in the Middle
East. Then, a result of increasing Mediterranean traffic, a unique though
relatively short-lived Minoan culture emerged on the island of Crete,
about 2000 BCE. These Bronze Age Cretans developed a unique system
of writing, knew how to make bronze tools, were skilled in the making
of wooden ships and construction of stone buildings, and were ruled by
kings with a considerable propensity for luxurious palaces. Archaeological excavations at these grand palaces showed that they even installed
a system of indoor plumbing and sewage discharge. This remarkably
tranquil civilization was partially destroyed by a severe earthquake and
eventually came under the control of the emerging Greek civilization.

Subsequently and at various other times between about 2000 BCE
and 500 BCE, empires arose and vanished in the Middle East, including Babylonia, Assyria, Persia, and others. This led to some mingling of
cultures and eventually an expanding trade especially along the shores
of the Mediterranean, aided by the invention of coinage by the Hittite
people residing in present-day eastern Turkey. These political and commercial developments required increasing engineering activity such as

the following:

(a) Building of roads, housing, and protective city perimeters
(b) Construction of larger triangular-sail and square-sail ships
(c) Extension of sea ports and harbor facilities
(d) Provisions for road access and port storage

Of considerable importance in these developments was the emergence of the barreled windlass[†] for hoisting and pulling heavy loads; the subsequent development of the pulley, ratcheted winch, and gears made for a most useful family of multi-purpose devices.

From the Middle East and the island of Crete, the political, economic, and religious center of prominence tended toward the mainland and nearby coastal areas. This ensuing Greek civilization consisted of many small loosely related city-states which shared a similar language and mythology. It was in this milieu that several important engineering developments emerged:

(a) Widespread iron smelting now replaced bronze for tools and weapons, and thereby ushered in the Iron Age
(b) Establishment of a powerful Greek navy with vessels characterized by single-sail, multi-tiered rowers, and a fore ram for piercing enemy ships
(c) Emergence of aesthetically appealing temples, open-air theatres, marble and bronze sculptures, public squares, court yards, and impressive housing communities

The Greek civilization became particularly prominent beginning about 600 BCE and is recognized for its remarkable colonization, promotion of democracy, codification of geometry, and development of philosophical thought. Especially visible remains of Greek culture are their formal buildings with the Parthenon of Athens especially well known, begun in 447 BCE and completed only nine years later. This building is distinguished by superbly shaped columns and carved friezes, and also characterized by geometric proportions which have subsequently been widely adopted as formal architecture in the West, Fig. 3.4.

Natural philosophy — that is the use of rational thought in the pursuit of a coherent conceptualization of nature — had its stimulation with Greek thinkers such as Socrates (469–399 BCE), Plato (427–347 BCE),

---

[†]The windlass (also known as a capstan) — first noted about 1400 BCE in the Middle East — is often judged to be next in importance to the wheel.

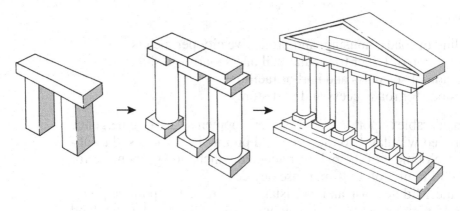

Fig. 3.4   Design evolution from prehistoric pillar and lintel, to Greek columns, and to Greek temple facades.

Aristotle (384–322 BCE), and others. The contemporary judgment is that, with some exceptions, their common avoidance of experimentation and disdain for manual work[†], introduced a bias about devices and some of their pronouncements and beliefs about the physical world have often proven to be ambiguous and frequently in error. However, among the prominent Greek engineers of the day to be noted is Eupalinos (∼500s BCE) for directing the construction of an extensive water supply systems requiring a 1 km tunnel, and Archimedes (287–212 BCE) for developing the principle of the lever, establishing the science of statics, discovering the law of buoyancy, and inventing the hollow helical screw for raising water.

Because of weakness from within and military pressures from without, beginning about 300 BCE, Greece declined in political importance. The Roman civilization — possessing an exceptional pool of military talent and effective means for producing professional soldiers and choosing skilled generals — then became increasingly dominant, eventually establishing its authority around the Mediterranean Sea and introducing significant advances in the continuing evolution of engineering.

To begin, Romans added a unique new feature to the practice of engineering: efficiency of large-scale and extended construction in support of military-political objectives. During their ∼600 years of domination over the Mediterranean region and western Europe, they built and maintained a network of over 100,000 km of all-weather roads linking some 4000 towns and cities. Their bridges and aqueducts proved to be

---

[†]Note that slavery was widely practised in Ancient times.

of remarkable endurance, many still standing as ruins and some even now still in use. Further, they built grand public structures such as the Colosseum in Rome holding 80,000 spectators — the largest structure of its kind in the world until the 1900s — and the Pantheon, the largest circular dome structure of its day. Additionally, the Romans built some very ingenious waterwheel-driven grist mills and some very effective human powered treadmill lift wheels for draining their copper, tin, and iron mines. Finally, they erected a diversity of walled defensive structures, improved on offensive and siege weaponry, and refined hand weapons and body armor.

A design feature invariably associated with Roman aqueducts is the semicircular arch composed of wedge-shaped stones so that gravity could hold the arch in place, Fig. 3.5. Adjacent arches and even tiers of arches were superimposed with stones tightly fitted so that mortar was not always necessary[†].

Fig. 3.5 Design evolution beginning with the Roman arch and leading to Roman viaduct/aqueduct structures of various length and multiple tiers.

Roman Engineers

Given so many Roman structures still standing throughout Europe, it is remarkable that so little is known about their builders. Some information has been culled from sparse historical records, including tombstones: Appius Claudius, builder of the famed Appian Way leading southward from Rome; Marcus Agrippa, chief waterworks engineer for Rome; Marcus Vitruvious, author of a pioneering textbook on design and construction methods. It is, however, also known that Roman engineers enjoyed considerable social status and prestige.

[†]The Romans did mix sand + lime + volcanic ash to obtain a good quality mortar and used it selectively. This recipe had become *lost* for about 800 years after the fall of the Roman Empire.

An appropriate primal heterogeneous progression which encapsules much of the important work of the Ancient engineers is suggested by the primal

$$N(t) \left\{ \begin{matrix} ore, \\ water, \\ stone \end{matrix} \right\} \rightarrow E(t) \left\{ \begin{matrix} design \\ project \\ organization \end{matrix} \right\} \rightarrow D(t) \left\{ \begin{matrix} mechanical \\ tools, arches, \\ roads, viaducts, \\ mortar, \\ waterwheels, \dots \end{matrix} \right\},$$

(3.8)

and encapsules also the superb capabilities of Roman engineers.

3.7 Tool/Device Development

Handheld tools were the first category of ingenious devices made. A vital development already incorporated in some of the earliest making of stone tools, was the practice of sequential changes towards their increasing utility. This was evidently an intuive process in all early devices. We may suggest this process by introducing a recursive loop in the primal (1.3) and (1.6), and which also appears as a specialization of (3.5), (3.7), and (3.8):

$$N(t) \rightarrow E(t) \left\{ \left(\begin{matrix} creative\ thought \\ skilled\ actions \end{matrix} \right) :\rightarrow \overset{\frown}{D'(t)} \rightarrow \right\} \rightarrow D(t). \quad (3.9)$$

Here $D'(t)$ is to suggest a provisional device suitable for direct use or it may serve as a test or prototype for further improvement by recursion.

Two important consequences to the evolution of engineering may now be noted.

3.7.1 *Material-Use Evolution*

An important and widely used time-interval labeling in history relates to the choices of materials used in making devices. Recall that tool making began about a million years ago using stone, wood, bone, sinew, and other animal and vegetable materials. The Stone Age ended about 3500 BCE with the discovery of bronze ($Cu + Sn$), an alloy which is shatterproof, holds a sharp edge, and is hard enough to provide an improved material for tools and other objects. This Bronze Age lasted about 2000 years, from about 3500 BCE to approximately 1500 BCE, when iron mining and smelting was discovered and became increasingly used. Iron smelting requires higher temperatures than copper and tin smelting, $\sim 1500°C$ compared to $\sim 1100°C$ and $\sim 250°C$ respectively, and was accomplished by using forced draft heating as well as coke and charcoal as the more dense fuel; these solid fuels were obtained by heating coal or wood in tight spaces so as to expel its trapped water and gases, yielded higher combustion temperatures. Iron then became a common metal most suitable for tools and other devices and thus identifies the Iron Age which lasted from about 1500 BCE until the discovery of large-scale steel making, beginning about 1860 CE. The correlation scale

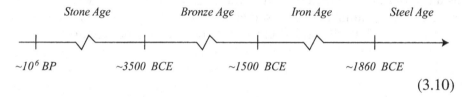

$$(3.10)$$

provides an overview of these Ages and also illustrates how materials for tool making are used to identify time coordinates and time intervals.

3.7.2 *Device-Use Fitness*

The recursive loop of Eq. (3.9) was initially employed by intuition long before it became formalized in classical Greek and Roman times. It clearly suggests that tool making involves successive changes towards greater utility and specialization. For example, the known variety of early stone tools illustrates their evolving speciation for emerging interests: chopping, cutting, scraping, pounding, spearing, etc. This successive development of early tool making alludes to the more general subsequent evolution of devices towards increasing *fitness*, much as the making of a key relates to its eventual fitting of a specific lock. In this sense, the development of devices $\rightarrow \overset{\frown}{D'(t)} \rightarrow D(t)$ possesses some features commonly associated with biological evolution.

3.8 Ancient Engineering: Societal Interest in Devices

This chapter has illustrated the continuing evolution of engineering now extended to the Ancient period ~ 8000 BCE to ~ 500 CE[†]. This period

---

[†]Appendix C provides a tabulation of inventions from this Ancient period.

is also significant because it marks the development of a broadly-based national interest in specific engineered devices: the various inventions by the Sumerians; the pyramids, temples, and ships by the Egyptians; the remarkable buildings, ornate stone sculptures, and exquisite weaponry by the Greeks; the very substantial roadways, aqueducts, and large public buildings by the Romans.

But aside from the authorization and management of these undertakings by some central authority, one may now also identify the first appearance of a societal receptivity in these technical initiatives. Hence, a broadly-based societal interest — to be designated by the symbol $S(t)$ — emerged as an important associated term to the engineering primal. We add that throughout all human history, societal interests $S(t)$ have evidently varied in detail but all have dominantly related to a perennial triad:

(a) *Political*
   Authority, security, boundaries, policy, warfare, justice, governance, ...
(b) *Economic*
   Subsistence, trade, labor, wealth, finance, manufacture, commerce, ...
(c) *Religious*
   Faith, morality, supernatural, spirituality, rituals, beliefs, worship, ...

Evidently it is possible to identify devices $D_i(t)$ which relate to specific aspects of these societal factors represented by $S(t)$.

In this Ancient period then, engineering advanced beyond Prehistoric Engineering, Sec. 2.7, establishing itself as the supplier of ingenious devices based on a perceived societal interest. This capacity to meet a societal interest established a corresponding Ancient Engineering connectivity progression characterized by extending the progression of the preceding Prehistoric Engineering era (2.4) now to read

$$N(t) \rightarrow E(t) \rightarrow D(t) \rightarrow S(t), \tag{3.11}$$

with $S(t)$ representing the set of relevant societal interests in and preferences for devices.

Table 3.2 depicts this evolution of engineering by extension of Table 2.1.

Table 3.2   Evolution of engineering from Prehistoric to Ancient times, $\sim$8000 BCE $\rightarrow$ $\sim$500 CE. Note that the Content Summary in this depiction represents a 2×4 matrix with variously interrelated and time-varying components. Further, our time axis is evidently nonlinear and will continue to be so in all subsequent depictions.

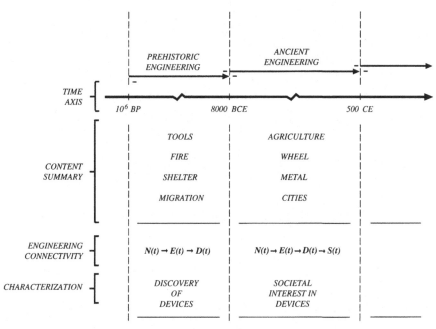

3.9   To Think About

- Archimedes was a most prolific inventor in Ancient times. Research some engineering details of his devices.
- Wood, stone, adobe brick, and mortar were the only building materials available to Ancient engineers. What distinctive additional building materials have since then been discovered or developed and are now commonly used?
- The invention of the wheel is remarkable because there exists no functional analogue in nature to be copied. Discuss the notion of invention by extension of observables and non-observables.

CHAPTER 4

Medieval Engineering

($\sim$500 CE $\rightarrow$ $\sim$1400)

Societal Promotion of Devices

4.1 Fragmented Landscape

Following the decline of the Roman military authority throughout Europe and the Mediterranean basin, traditionally dated to the late 400s CE, a patchwork of regional duchies and various types of fiefdoms appeared, many governed by self-appointed autocrats ruling over mostly subsistence farmers and crafters. Geographic boundaries were inadequately defined or selectively ignored and regional rivalries, border skirmishes, *barbarian* invasions, and shifting alliances were frequent. Roman roads and bridges soon fell into disrepair thus hampering trade and travel, and thereby tending to isolate towns and villages. Individual opportunity was at a low ebb with fatalism, alchemy, and astrology prevalent. Only a small fraction of the populace could read and write so that little more than a thin layer of culture could be said to exist in Medieval Europe[†].

Nevertheless, crafting pursuits and a search for material betterment could not be suppressed. Woolen clothing, because of its tendency to cause itching followed by scratching and irritation of the skin, became slowly replaced by linen clothing. Harvesting and storage of hay was introduced, making it possible to use horses further north and for extended periods of the year. Land drainage was continued thereby converting more northern wetlands and estuaries into productive agricultural lands. The ensuing agricultural surplus and continuing crafting developments combined to introduce a new and most important institution throughout Europe: the village market, a most important social and economic stimulant in Medieval times.

[†]Elsewhere, for example in the Viking and Islam dominated areas, this time period continued to be characterized by considerable vitality.

Further, a sense of private ownership slowly emerged leading to much interest in the making of labor-saving devices. Also, small-scale cottage enterprises developed providing for excess labor to engage in teaching and learning. Now and then coordinated interests in larger engineering projects developed, related generally to church, military, and aristrocracy interests. Slowly, Medieval Europe came to life.

4.2   Hagia Sophia

A singular and most remarkable engineering construction project in Europe can be traced to the early Medieval period. As part of the allocation of some religious and political authority from Rome to the East, there emerged an interest to establish an appropriate and impressive ecclesiastical building. The ancient Greek city of Byzantium was chosen as the site for this initiative and renamed Constantinople — and in 1453 renamed again as Istanbul. What resulted about 530 CE with the employment of some 10,000 craftsmen and in the short span of only 5 years, was an architectural and engineering triumph: a large square 37 m × 37 m × 30 m basilica was for the first time satisfactorily topped by a large dome, Fig. 4.1. What made this building construction advance possible was the

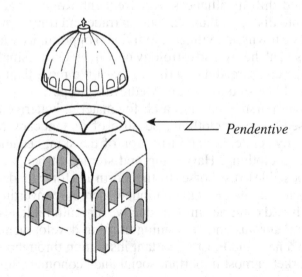

Pendentive

Fig. 4.1   Architectural design concept for the placement of a dome on a square structure: Hagia Sophia, ~530 CE. Large naves and aisles directed outward formed an extended complex of connected partially open vaulted spaces. These outward extensions also provided stability to the main walls holding up the dome.

placement of ingenious curved-triangular sections between the rounded tops of the main walls — the so-called pendentives — thereby forming a circular base for a large dome. As an added feature, arched windows were placed around the base of the dome so that when viewed from the inside the dome appeared to be floating. This magnificent basilica was named *Hagia Sophia*, Greek for Holy Wisdom, and remained the largest religious building for over 1000 years.

Impressions of Exceptional Buildings

Buildings may overwhelm and inspire. At the dedication of Hagia Sophia, Byzantine Emperor Justinian I — under whose sponsorship it was built — is said to have been so moved by its grandeur that he exclaimed "Solomon, I have surpassed you", a reference to the Hebrew King Solomon and his building of the Temple in Jerusalem some 1500 years earlier; Hagia Sophia, sometimes also referred to as Santa Sophia, has inspired architects and structural designers ever since. Four large slender minarets were later added in keeping with its ensuing function as a mosque. It presently serves as a museum and Istanbul's most prominent tourist attraction.

4.3   From Agriculture to Power Plants

By about 600 CE, a device slowly appeared in central Europe with very important consequences: the wheeled iron plow drawn by a pair of oxen, Fig. 4.2. Recall that agriculture in the tropical and sub-tropical region surrounding the Mediterranean involved loose soil for which the shallow wooden scratch plow was adequate. Europe, however, was heavily forested and covered with loamy soil for which the scratch plow was totally inadequate. The iron plow cut deeper, ~ 15 to $25\,\text{cm}$, while its mouldboard inverted continuous strips of soil, thus mixing and loosening it for seeding. This engineered device stimulated the following significant innovations throughout Europe:

(a) Increase in arable land, adaptive selection of plants, and introduction of crop rotation.
(b) Crafting activity initiatives specifically related to harvesting tools, leather tanning, cart and wagon construction, and various other implements.

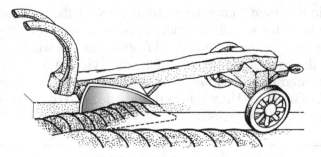

Fig. 4.2   The wheeled iron plow of northern Europe, introduced about 600 CE. Its unique feature was the iron mouldboard permitting the inversion and loosening of the top layer of soil.

In addition to agricultural developments, other inventions and adaptations occurred — sporadically and intermittently and with substantial variations in space and time. Among these we cite the vertical axis cloth-braced windmill in the area of present-day Iran, about 700 CE. Techniques of ink block printing in which characters stood out in relief on wood or stone surfaces, slowly diffused from China ($\sim$200 CE) to Korea ($\sim$600 CE), reaching Japan about 800 CE. The important crank handle for turning an axle — possibly first used in China $\sim$200 BCE — became widely applied in Europe beginning $\sim$800 CE. And, everywhere, crafters were producing ingenious tools for working with wood, stone, and metal.

For skilled and ambitious crafters, the Medieval period proved to be an important gestation opportunity, ultimately leading to innumerable adaptations and variations of the so-called Greek *Fundamental Simple Machines*: lever, wedge, wheel and axle, screw, and pulley. Among the more practical innovative uses and refinements ensuing from these basic device types, we list additionally the windlass, winch, crane, hoist, block-and-tackle, crowbar, bevel gear, pump handle, pulley drive, worm gear, shipwheel, ratchet wheel, ...

Two devices also appeared in Europe with especially significant consequences: the iron horseshoe, $\sim$900 CE, and the horsecollar, $\sim$1000 CE[†]. The former enabled horses to travel on stony ground and on the expanding gravel and cobblestone roads, and the latter allowed horses to pull a wagon or plow without choking.

[†]There are indications that crude forms of both horseshoe and horsecollar may have first been used in Asia, $\sim$100 BCE.

Incorporation of the horse into human affairs provided another power source for transportation. Recall that until this time only humans and cattle were available to provide speed and power. Average values for these two engineering parameters can be cast into the progression

$$\begin{pmatrix} Humans: \\ \sim 5\,km/hr \\ \sim 100\,W \end{pmatrix} \rightarrow \begin{pmatrix} Cattle: \\ \sim 2\,km/hr \\ \sim 400\,W \end{pmatrix} \rightarrow \begin{pmatrix} Horse: \\ \sim 6\,km/hr \\ \sim 800\,W \end{pmatrix}. \tag{4.1}$$

Meanwhile, medieval monks continued to labor diligently in copying books, recording astronomical phenomena, and documenting political events, while also improving methods for communal self-sufficiency. In the secular world, village markets considerably aided social contact, commerce, food marketing, and the crafting trades. In the political realm, Emperor Charlemagne established schools about 800 CE and encouraged education throughout Western Europe. Eventually scholars began to assemble and the first universities were founded in the mid 1100s CE, initially devoted to the teaching of theology and training of clergy, and later to the study of medicine and law; the universities in Bologna, Oxford, and Paris were the first such European institutions, dating their embryonic origins to Medieval times.

The centuries immediately following the Roman decline included slow changes in which new devices were important enabling tools. These changes may well be represented by the heterogeneous progression

$$\begin{pmatrix} disorganization, \\ poverty \end{pmatrix} \rightarrow \begin{pmatrix} crafting, \\ commerce \end{pmatrix} \rightarrow \begin{pmatrix} church\text{-}based \\ education \end{pmatrix} \rightarrow \begin{pmatrix} organization, \\ scholarship \end{pmatrix}.$$

$$\tag{4.2}$$

Underlying much of this progression — and tracing its origin to early Roman times — was the continuing and remarkable development of the waterwheel. With rain water scarce in the Middle East, primitive wheels with tilting pails along the rim were used to raise river water for irrigation. In Europe, with rain more plentiful and snowmelt also contributing to cascading streams, the ingenious Roman engineers developed a suitable bevel gear and associated structures to render the waterwheel a more useful rotational power source; subsequently, clever Medieval engineers extended the axles of waterwheels and mounted cams to provide a useful pulsed power source. Two hydro-energy applications thus became increasingly specialized and important:

(a) Power for grinding (introduced ~100 CE): milling grain, grinding stone, honing tools, polishing brass, . . .
(b) Power for trip-hammering (begun ~1000 CE): crushing ore, forging metal, stamping coins, pulping rags for paper, tanning leather, operating blacksmiths bellows and foundry blast furnaces, . . .

Both of these classifications provided much opportunity for mechanically minded and creative individuals.

Power from Water

Experience over the millennia contributed to a recognition of water also as a profound source of power. First, the people of Mesopotamia used the natural sloping water flow for irrigation; then, Roman engineers expertly converted this linear water motion into rotational motion for grinding wheat; in Medieval times, unknown engineers mounted cams to the extended axles of water wheels thus inventing the triphammer. The heterogeneous power progression has continued:

$$
\begin{pmatrix} linear \\ power \\ \sim 4000\ BCE \end{pmatrix} \rightarrow \begin{pmatrix} rotational \\ power \\ \sim 100\ CE \end{pmatrix} \rightarrow \begin{pmatrix} pulsed \\ power \\ \sim 1000\ CE \end{pmatrix} \rightarrow \begin{pmatrix} steam \\ power \\ \sim 1712 \end{pmatrix} \rightarrow \begin{pmatrix} hydroelectric \\ power \\ \sim 1900 \end{pmatrix}
$$

$$(4.3)$$

Thus, water sustains not only biological life but also industrial life.

4.4   Developments in Isolation

In addition to the ancient population clusters in the Middle East, a separate population center developed about 2000 BCE along the Yellow River of present-day northern China. These people became technologically productive under a unique set of political and cultural circumstances.

To begin, a deliberate policy of superior isolation had long been a characteristic pursued by a succession of ruling Chinese dynasties. This included the construction of a tangle of protective barriers along its northern borders. Beginning about 200 BCE the various sections were organized into a linked fortification with construction of new sections occurring intermittently until about 1600; the eventual length of this defensive rampart approached 4000 km and is commonly called the Great Wall. Another extensive construction project was the 2500 km Grand Canal south from Beijing and connecting their eastern cities with an effective trade route. Constructed also intermittently from ∼100 CE to ∼1300, it included some 40,000 lakes and — most remarkably — it crossed a substantial mountain range requiring very sturdy slipways, an independent lubricating water source at each divide[†], and a powerful windlass system for moving boats over each summit. More than 5 million workers were involved in this largest ever hydraulic engineering undertaking.

In general, the Chinese leadership and bureaucracy provided an environment for considerable inventive effort, resulting in the development of a number of unique devices some of which eventually proved to be of global importance: from Ancient times there appeared ink, lodestone magnet, and paper, while Medieval times provided porcelain, and gunpowder. However, because of its restrictive and condescending policy towards external contacts, many of its inventions and innovations did not diffuse to the wider world for centuries.

Some specific cases of delayed cross border diffusion of processes and devices can be identified. It has been established that while paper in the form of compressed felted vegetable fibers was first made in China about ∼200 BCE, it took 400 years for it to appear in Japan and another 900 years to arrive in Europe; note that the predecessor to paper was papyrus (obtained from pounding cross-layered slices of reed) and parchment (obtained from goat and sheep sinew and skin), both first produced in Ancient Egypt. Further, high quality porcelain was manufactured in China beginning 500 CE and became highly prized by the European aristocracy about 300 years later; eventually, by about 1600, suitable ore deposits were found in Eastern Europe and techniques developed for comparable quality porcelain manufacture.

Explosive incendiary mixtures are another profound Chinese invention though the detailed early development continues to be unclear.

[†]Generally obtained by diversion of nearby higher-elevation streams of the watershed.

It appears that a volume composition consisting of ~20% charcoal, ~20% sulfur, and ~60% potassium nitrate, had been initially used by the Chinese as a noise maker or firecracker about 600 CE and later, about 1100 CE, also used to power crude bamboo rockets in a conflict with Mongolia; Europeans first used gunpowder in the mid 1300s.

Chinese records of ~800 BCE refer to lodestone (i.e. magnetite = magnetic oxide of iron) attracting iron chips. But it was not until about 1000 CE, that a unique direction pointing device became known in China: a small fish-shaped sliver of iron — ceremoniously magnetized by stroking with lodestone — when floated in a dish of water would align itself in a North-South direction. This property did not become widely known in Europe until about 1200. Dry-mounted versions as a boxed balanced magnetized needle with a 32-direction wind rose as background began appearing about 1300 and proving to be of critical importance to the ensuing global exploration by seafaring Europeans.

As a contrast to the above outward migration of Chinese device development, it is to be noted that the two-wheeled chariot first appeared in Sumer about 2500 BCE but its use in China did not occur until about 100 BCE. Also, glass had been first manufactured in Egypt about 2000 BCE, but was not produced in China until ~600 CE.

Evidently, knowledge and techniques of the making of ingenious devices had eventually become more mobile, passing from one regional culture to another. Important in the East/West diffusion was the so-called Silk Road — the several land trade routes joining Europe, Middle East, South Russia, India, and China — serving as a trade corridor for selected devices and agrarian products, beginning ~500 BCE. Flowing East-to-West was primarily silk as well as spices (e.g. pepper, cinnamon, curry, ...), tea, textile dyes, and porcelain; flowing West-to-East were metals, furs, vegetable seeds, and glass products — and eventually gold in payment. No trader traveled the entire Silk Road so that goods were repeatedly bartered and passed through many hands before finally ending up at faraway destinations.

And, even though inventions and device development occurred at different times and in different regions, the shared primal

$$N(t) \rightarrow E(t) \rightarrow D(t) \left\{ \begin{array}{l} \textit{local-regional use} \\ \textit{long-distance trade} \end{array} \right\} \qquad (4.4)$$

could be identified as universal.

Medieval Times in America

Beginning about 600 CE the Anasazi people of the North American Southwest developed a unique culture. They constructed impressive interconnected stone and masonry dwellings (called *Pueblos* by the later Spaniards) on mesas and buttes, in canyons, and under overhanging cliffs. As an exceptional example, the Pueblo Bonito community located in the area of present New Mexico, contained ~600 housing units, some up to 4 stories for which wooden beams had to be transported about 100 km by human labor (no wheel, no draft animals); they practiced dry farming (corn, beans, squash), hunted with bow and arrow, had developed painted pottery, crafted turquoise ornaments, and conducted rituals in below-grade circular *kivas*. For unknown reasons, these buildings were deserted about 1400. The Maya in Central America were also thriving during these Medieval times while the Aztec, also in Central America, and the Inca in South America flourished later, about the 1400s.

4.5 Religious Influence

Beginning about 630 CE, a profound religious and cultural change occurred in the Middle East and surrounding regions. The people of Arabia became inspired by the teachings of the Prophet Muhammad (570–632 CE) and within a century Islamic thought and practice had spread north to Asia Minor, east to India, south to northern Africa and from there to Spain. Adopting also a newly found passion for learning, Islamic scholars collected, translated, transmitted, and contributed to the knowledge of the Medieval World; indeed, it was through their translations that Europe learned more about Classical Greece. From about 700 CE to 1200, Islamic technology exceeded that of Medieval Europe. In particular, their medical skills were the best available and their libraries were most extensive. Additionally, Islamic ability to breed spirited horses became known and their metallurgical skills were widely recognized with the *Damascus* sword, patterned carpets, and decorative copper wares especially prized.

Of particular interest to engineering is that Islam actively promoted the standardization of weights and measures. They also chose to use the Indian methods of numerics, beginning about 800 CE; Europeans did not adopt it until about 1200. The most significant aspects of Indian numerics consisted of the following:

(a) Ten distinct number symbols, including zero (now commonly represented by 0, 1, 2, 3, . . . , 9) and noting that the number zero possessed some exceptional properties.
(b) Decimal place-value notation for multidigit numbers so that one could distinguish, for example, the number 3096 from 3906.

This Hindu-Arabic method proved to be a significant improvement over Roman numerics.

Arabic Spoken Today

The Arabic mathematician *Muhammed ibn Musa al-Khwarizmi* (~780–850 CE) wrote a book on practical mathematics titled *al-Jabr* in which Indian numerics were used throughout. By a rough transliteration, the title of this book led to the English word *algebra* and the author's surname name is perpetuated by the term *algorithm*. Additionally, the medieval European practice of seeking the extraction of precious materials by distillation has its origin in Egypt as *El-khem*, subsequently giving Europe the word *alchemy*.

Islam began building mosques first throughout the Middle East and subsequently elsewhere. These sacred places have a central dome with interior geometrical pattern wall decorations, carpeted floors, wash basins at the entrance, a recessed niche facing Mecca, and slender minarets containing balconies from which the faithful are called to prayer.

Construction of large Christian sacred places in central and western Europe began about 1150 with three significant structural innovations enabling the construction of tall and visually impressive cathedrals:

(a) Pointed arches as extensions of the Roman circular arch
(b) Flying buttresses allowing taller walls by load distribution
(c) Ribbed ceilings permitting longer structures

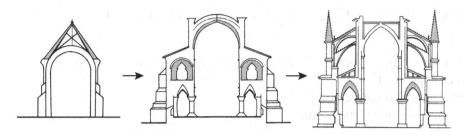

Fig. 4.3   Evolution of the Gothic cathedral designs leading to flying buttresses. Some cathedrals were also characterized by tall pointed spires. (Not to scale)

By 1250 some 250 large and graceful cathedrals had been started in Western Europe, Fig. 4.3. Thus,

$$N(t) \begin{Bmatrix} stone, glass, \\ wood, brick \end{Bmatrix} \rightarrow E(t) \begin{Bmatrix} graceful\ designs \\ complex\ construction \end{Bmatrix} \rightarrow D(t) \begin{Bmatrix} sacred \\ buildings \end{Bmatrix}$$

(4.5)

characterizes the primal progression associated with the Gothic style of sacred structures throughout Christendom, requiring however long construction times — on time scales of centuries — and involving exceptional scaffolds, extended winches, and superior crafting skills. Inclusion of colored stained glass windows, elaborate altars, and ornate stone and wood carvings added considerably to their public appeal. Churches with tall spires often contained large bronze bells which were used to call the faithful to worship.

The common label for the skilled designers and supervisors of these novel structures was master mason. Though often illiterate, these experienced and skilled individuals acted very much like present-day consulting engineers or project managers, some also becoming influential and in high demand. Nevertheless, an indication of the trial-and-error building methods of the day is the record of partial collapse of some cathedrals during construction.

4.6   Military Influence

Greek and Roman military crafting practices had provided for considerable metal forming skills useful for body armor manufacture as well as for walled fortifications and siege weaponry such as catapults. And then, beginning about 800 CE, a new form of offensive military action of the Medieval period emerged: the raids of the Vikings along the oceanic shores and inland rivers of Europe. What made these forays technically possible were several unique naval innovations:

(a) Wide shallow-draft boats of overlapping hull-boards commonly called clinker-built
(b) Multiple oars and single square sail
(c) Aft steering for one-person control

This maritime device led to some significant social and political consequences. For example, while many Viking raids were predatory, others were known to lead to Viking settlements in their lands of conquest — including England, Iceland, northern France and western Russia. Indeed, the first short-term European settlers in North America were Vikings who landed in Newfoundland about 1000 CE arriving from Iceland via Greenland. By ~1100, the Viking raids had run their course.

Further, a small but highly consequential device emerged in Europe during the middle of the Medieval era: the metal stirrup[†]. This small, innocuous device offered great offensive military advantage to a skilled rider with a sword: speed, surprise, powerful swinging action, and fast escape. It proved decisive in some battles, most notably in the Norman conquest of England in 1066. In time, the rider became equipped with a shield and other armor, adding considerably to the power and fear of this kind of military action. This simple stirrup device was soon instrumental in establishing the medieval European knighthood tradition and subsequently introduced the cavalry as an important component of the military, the latter surviving even into the early 1900s.

Then, while early defensive structures were associated with strengthened and expanded dwellings of a regional duke or prince, strategic military considerations soon led to expanded requirements of engineering relevance. It began with moats and thick stone wall structures, designed also to protect exploitative rulers from a restless or angry populace. Fortification developed by ~1100 CE into elaborate turreted castles of elevated thick-walled stone and masonry defense installations surrounded by moat and drawbridge. In some cases, defensive walls surrounded entire towns. Also, large ingeniously designed fortresses emerged equipped with hidden passages and storage spaces containing extensive supplies

---

[†]While the padded saddle, rope, rein, and leather stirrup may well have been used in the steppe of Asia for centuries, the rigid metallic stirrup did not appear until about 200 BCE in India. This most important device first came to be widely used in the Ukraine beginning about 500 CE.

to withstand a long siege. And so the primal progression

$$N(t) \begin{Bmatrix} stone, \\ brick, \\ iron \end{Bmatrix} \rightarrow E(t) \begin{Bmatrix} dynamic\ design \\ heavy\ construction \end{Bmatrix} \rightarrow D(t)\ \{fortifications\}$$

$$(4.6)$$

provided for aspects of military engineering of Medieval times.

Events which combined religious devotion with military zeal also resulted in some important consequences. It so happened that the remarkable agricultural productivity of the Medieval feudal system in Europe soon required fewer farmers so that an overproduction of knights occurred with many eager to take on the Papal challenge of recapturing Jerusalem from Islam. The consequent pseudo-military expeditions during the ~1100 to ~1300 period did not ultimately succeed but the associated exposure to Byzantine and Islamic culture did lead to a broadening outlook in Europe, including an interest in Eastern practices and devices. Most significantly, the returning crusaders brought with them familiarity with incendiary devices which had previously been brought from China to the Middle East. European craftsman improved on these explosives and developed crude cannons first used in 1346 and in muskets soon after.

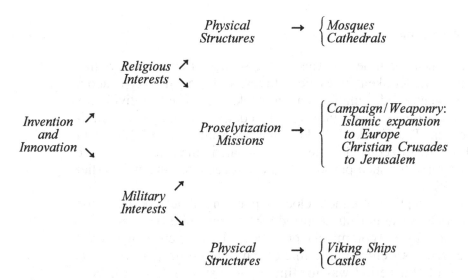

Fig. 4.4 Connectivity associated with religious and military influence of Medieval time.

One may notice the complementary convergence between religious and military interests of the late Medieval period as beneficiaries of the engineer's inventive capacity. Obviously, both interests relied upon the technical knowledge and skill of the engineers of the time and both have an interest is physical structures, Fig. 4.4.

Medieval Device Diversity

Agricultural practices, religious interests, military objectives, and commercial activities of medieval times all influenced crafting of the day and thus contributed to an extension of heterogeneous progressions as suggested in the following:

$$\begin{pmatrix} stone,\ wood, \\ metal,\ fiber, \\ leather, \\ minerals \end{pmatrix} \rightarrow \begin{pmatrix} selection, \\ preparation, \\ shaping, \\ assembly \end{pmatrix} \rightarrow \begin{pmatrix} various \\ and \\ numerous \\ devices \end{pmatrix} \rightarrow \begin{pmatrix} agricultural\ interests, \\ religious\ interests, \\ military\ interests, \\ commercial\ interests \end{pmatrix}.$$

(4.7)

4.7 Timekeeping

Attempts to measure time intervals on the scale of fractional days had already been undertaken in Ancient times. It began in Egypt about 3000 BCE when crude sundial shadow sticks were used to divide the time from sunrise to sunset into 8 or 12 equal time intervals, regardless of season. The Greeks introduced cumbersome waterclocks about 400 BCE in order to control the length of debate in their Senate; the use of the hour glass for such purposes had also been cited in some earlier Greek writings.

With the exception of candle clocks appearing about 200 CE, little change in time measurements occurred for the next 1000 years. By then, monastery activity and commercial practices led to an increasing interest for a more precise specification of time-of-day. Interestingly, the common expectation of that period was for time coordinates to be heard, as in pealing bells, and not necessarily seen as a pointer referring to a number near the rim of a disk.

By the early 1300s a profound intellectual and crafting fusion became established among some clever south European craftsmen: it seems to have become intuitively evident that gravity could possibly serve the purpose of sustaining a force for generating vertical distance intervals Δs as proxy for time intervals Δt so that adding a number of these intervals would then yield a desired time coordinate t. And so — now using iron crafting techniques — the challenge then became the capturing of repeatable Δs intervals in some autonomous device. The idea of a weight attached by rope to a frictionally controlled rotating drum (i.e. a windlass) must have been repeatedly tried but evidently found futile because acceleration could not be eliminated. But then, by the genius of one or several inventive minds — nobody knows who or where — the remarkable verge-foliot control device appeared about 1350; in this escapement mechanism, a constant Δt became associated with a weight moving by force of gravity a constant vertical downward distance Δs thereby generating a repeating force required to impart a specific horizontal rotational energy of the foliot pivoting on the verge, Fig. 4.5. The time required

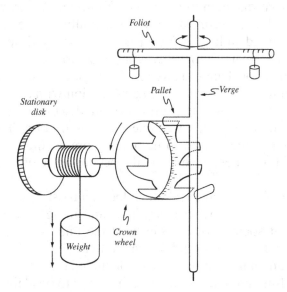

Fig. 4.5 Isometric and operational depiction of the verge-foliot regulated clock: the verge is forced through back-and-forth swings by contact with the crown wheel; the corresponding swing action of the foliot involves a time interval Δt, specific to the mass and separation distance of the balance weights; the weight on the rotating drum insures repeated and identical Δt with each swing action advancing a gear arrangement on the stationary disk thereby moving a hour-hand, ultimately yielding a 360° rotation in a noon-to-midnight period.

for this transfer of energy yielded the constant Δt. This design concept proved to be sufficiently robust and many such time pieces — with many variations — were constructed.

Large clocks were first mounted in specially designed clock towers as community showpieces and eventually installed in church steeples. Bells were installed and suitable bell-ringing mechanisms attached to peal at $\sim 1/24$ of a day-night cycle. An hour indicator only was first used since the inherent inaccuracy rendered a minute hand pointless. The primal progression descriptive of this engineering developments is suggested by

$$N(t) \{metals, \, gravity\} \rightarrow E(t) \{gears, \, kinetics\} \rightarrow D(t) \{gravity \, clock\}.$$
$$(4.8)$$

The verge-foliot escapement clock — also known as the weight-driven or gravity-driven clock — is widely recognized as a most profound creative achievement of Medieval times; additionally, it was also the first substantial-size device made entirely of iron. Numerous improvements soon emerged and the increasing accuracy eventually justified the installation of a minute-hand indicator. Nevertheless, improved accuracy, size reduction, and greater reliability continued to be sought. It was not until 1650 that the Dutch astronomer Christiaan Huygens adopted Galileo's pendulum swing dynamic in conjunction with a novel pendulum-triggered escapement — but with a weighted windlass still required to provide the motive force to compensate for friction. This pendulum-controlled time piece proved to be much more accurate and smaller than the verge-foliot design, eventually becoming a prestige household show piece.

The wide acceptance of the mechanical clock led to a most profound change in social and intellectual perspective. To some it signaled the eventual regimentation of society; others saw in it the understandable function of a synchronized system — and hence a remarkable model of the operation of many human institutions; some concluded that since the solar system seemed to function like clockwork, there had to be a divine clockmaker.

4.8   Consequences of Wind and Water

Wind and water, or to be more accurate, blowing air and flowing water, had particular significance for Medieval Engineering because these natural phenomena had the capacity to exert a force and this force could substitute for human effort. Already Ancient Engineers were aware of the effects of flowing water in their construction of irrigation channels as well as in the effect of wind on sail-equipped water crafts. Additionally, moving air has long possessed a spiritual connotation since many ancient languages use the same word for *soul* and *life* as for wind.

Clock Conventions

The introduction of the verge-foliot clock serves as an example of the frequent need for convention or protocol whenever some new device is introduced. Until then, time periods had names such as First Watch, Second Watch, etc. as determined by the sun and an hour glass. Further, with clock designs providing for a circular scale and 12 daylight time intervals also becoming common, some early bell tower clocks possessed 24-hour circumference markings and others had pointers rotating *counterclockwise* on the clockface. However, *clockwise* motion on a 12-hour disk eventually became the standard. Then, in 1883, a major global protocol became established: Universal Standard Time consisting of 24 one-hour global longitudinal regions referenced to the 0° Prime Meridian passing through Greenwich, an observatory in England.

It was known to the Greeks, and it must have been recognized by many in Medieval times, that the application of wind and flowing water represented the first and extensive use of motive power beyond that available from animals or humans. Wind and flowing water initiated two distinct heterogeneous progressions which, interestingly, converged to provide equivalent functions in Medieval times:

$$
\begin{aligned}
&\textit{blowing air} \rightarrow \quad \textit{windmills} \\
&\qquad\qquad\qquad \begin{pmatrix} \textit{cloth, wood} \\ \leq 3000\ W \end{pmatrix} \searrow \\
&\qquad\qquad\qquad\qquad\qquad \begin{cases} \textit{agriculture: grinding, pumping, } \ldots \\ \textit{crafting: spinning, triphammering, } \ldots \end{cases} \\
&\textit{flowing water} \rightarrow \quad \textit{waterwheels} \qquad\qquad \nearrow \\
&\qquad\qquad\qquad \begin{pmatrix} \textit{overshot, undershot} \\ \leq 3000\ W \end{pmatrix}
\end{aligned}
$$

$$(4.9)$$

These two devices, windmills and waterwheels, provided much opportunity for technical invention and innovation during the Medieval period and this for various reasons. Each geographical site for such installations was unique and redirection of rotational motion involved various design choices; the former tended to involve geological and climatological considerations while the latter tended to require unique construction and mechanical skills. We add that the essential bevel gear, enabling a 90° change in rotational motion, was known since Roman times.

It is in the emerging area of coastal commercial shipping that wind and water combined to stimulate a human response for particularly impressive device developments. Already in the 1100s, Viking ships were redesigned and constructed with deeper holds for cargo. Then several important shipbuilding innovations appeared:

(a) Stern-post rudder centrally placed aft
(b) Bowsprit forward spar for sail extension
(c) Rear castle for the captain and steersman

The ship which thus appeared, typically a tubby ~25 m long with a ~8 m beam, was called a *Cog* and soon proved to be a most reliable workhorse for coastal trade of northern Europe, sufficiently so that by the mid 1200s a first supranational organization appeared — The Hanseatic League, a federation of about 100 coastal and riverine cities sharing a common interest in trade and commercial development. This development further stimulated crafting and agricultural development not only along the northern European coast but also connecting with Venice — an emerging staging point for trade with the East. The engineering and seafaring skill so established provided for increasing coastal exploration, eventually leading to sea travel along the west coast of Africa beginning in the mid 1300s.

And so the exploitation of wind and water initiated the divergence and specialization of device development in support of economical and political interests:

$$
\text{wind and water}
\begin{cases}
\nearrow & \begin{array}{l}\textit{sailing} \\ \textit{ships}\end{array} \rightarrow \begin{cases}\textit{transportation (trade, information)} \\ \textit{exploration (coastal, oceanic)} \\ \textit{military (defensive, offensive)}\end{cases} \\
\\
\searrow & \begin{array}{l}\textit{mills,} \\ \textit{triphammer}\end{array} \rightarrow \begin{cases}\textit{crafting} \ldots \\ \textit{agriculture} \ldots \\ \textit{mining} \ldots\end{cases}
\end{cases}
$$

$$(4.10)$$

Innumerable and distinct Medieval developments may be associated with such heterogeneous progressions: trade $\rightarrow$ banking, crafting $\rightarrow$ shipbuilding, transportation $\rightarrow$ insurance, exploration $\rightarrow$ national economies, naval strength $\rightarrow$ political authority, and others.

Medieval engineering, though initially associated with an inactive and lackluster part of human history, now became alive and of considerable importance in the continuing evolution of ingenious devices of relevance to a variety of commercial institutions.

A Medieval Catastrophe

By the early 1300s, the combined occurrence of (a) an earlier over-population leading to marginal land occupancy, (b) followed by climatological cooling (Little Ice Age) and hence years of reduced harvests, and (c) the consequence of regional famine leading to weakened health, rendered Europe weakened to the spread of pasteurella pestic bacterial infections — often called the Black Death epidemic. It started with openly decaying human bodies in Asia Minor and was transported by infected rats along the Mediterranean trade route to western Europe. The bacteria was transferred to humans by fleas which had bitten infected rats. As a consequence, premature death was widespread so that one hundred years later, the population of Europe was only about half of what it had been when the plague first struck. For this and other reasons, the 14th century is generally viewed as one of exceptional and tragic hardship for Europeans.

4.9   Changing Image of Engineering

In the discussions of Prehistoric Engineering, it had been noted that the engineer could be considered some skillful stone-knapper; in Sumerian times the engineer was a priest-builder and in Egyptian times he was a governor-builder; during the Greek era, the engineer could be equated with an architecton-designer and in the Roman era the ingeniator was the inventor of ingenious structures; finally, in the present Medieval era, the master-mason was the engineer-technologist well able to respond to a societal interest in a variety of sacred buildings. Hence, the evolution of labels for engineers can be cast in the format of the heterogeneous progression

$$\begin{pmatrix} stone \\ knapper \end{pmatrix} \rightarrow \begin{pmatrix} priest \\ builder \end{pmatrix} \rightarrow \begin{pmatrix} governor \\ builder \end{pmatrix} \rightarrow \begin{pmatrix} architecton \\ designer \end{pmatrix} \rightarrow \begin{pmatrix} ingeniator \\ inventor \end{pmatrix} \rightarrow \begin{pmatrix} mason \\ technologist \end{pmatrix}$$

$$(4.11)$$

thereby also suggesting a kind of evolving engineering practitioner.

Reflecting on these various characterizations of the engineer of Medieval times suggests the evolving identification of technically oriented individuals as principal contributors to meeting particular interests of society of the day. And with it came a public recognition of the value of skillful actions and creative thought of individuals.

While the engineers of the preceding Ancient era might well be involved with large projects based on the demands of a powerful religious or military authority, the late Medieval period introduced the increasing role of a broadly based direct societal support, thereby stimulating an expanded promotion for engineering undertakings. Examples are popular support for Gothic cathedrals, township support for clock towers, crafters' support of *Cog* shipping, commercial support of roads and bridges, farmers' support of implement manufacture, etc. Notwithstanding the implied role of special interests in specific engineering projects, one could now begin to recognize a broad societal influence affecting engineering in general and specific engineers directly.

Then, another factor appeared which subtly affected the practice of engineering in Medieval times: intuitive insight into the inner workings of devices. For example, already during the embryonic development of the gravity-driven clock, there emerged a focus on the intricacies of gears eventually yielding the family of remarkable astronomical clocks. Then, the Greek *Fundamental Simple Machines*, especially the lever and wedge, became integrated with circular motions to yield the screw jack — the forerunner of devices such as the wood screw, the metallic bolt, and the ship propeller. Also, continuing developments in crafting led to the complex horizontal loom and hence prospects for colorful textile patterns. And, at another level of thought, while fatalistic and pessimistic views of human life were still dominant, there now appeared writings on a positive prospect for society — aided by ingenious devices.

What is observed here is the identification of three factors associated with the continuing evolution of engineering:

(a) Recognition of special skills of artisans and crafters
(b) Recognition of the expanded importance of ingenious devices
(c) Recognition that both the engineers and their devices seemed to contribute — in various ways — to the idea of improving living conditions

The sum effect of these consideration, is not only an explicit approval by society of the works of Medieval engineers, but also an implicit encouragement in the continuing development of improved specific devices such as waterwheels, taller cathedrals, more visible clocktowers, labor saving implements made of iron, etc. This approval and encouragement can be taken to be a positive feedback emanating from society to the community of engineers of the day so that symbolically we may write the connectivity

$$N(t) \longrightarrow E(t) \longrightarrow D(t) \longrightarrow S(t) \qquad (4.12)$$

positive feedback

implying thereby a cyclical effect[†].

Maintaining our previous nomenclature progression, we see in $E(t)$ an expanded function of the Medieval engineer. And, recalling the analysis of Chapter 1, we now also note the emergence of an adaptive connectivity — that is a heterogeneous progression now reflecting a societal influence on the theory and practice of engineering. This is the beginning of an important concept: in the process of engineering influencing society through their ingenious devices, engineering activity is also unavoidably influenced by selective interests of society.

Small-Scale Devices

While engineering of Medieval times was evidently characterized by impressive highly visible large devices, some small device developments also proved to be most important. One particularly remarkable case involved the invention of glass lenses and eye glasses. The discovery of the visual benefit of convex glass disks occurred in northern Italy during the late 1200s and though the quality of glass was low, it did provide effective magnification. This basic consumer device substantially increased the working life of crafters at the peak of their skill. Not only was this a direct benefit to individuals and hence society, it also prepared the way for the subsequent development of microscopes and telescopes.

[†]Appendix D provides clarifying illustrations of combined operational mappings and cyclical representations.

4.10   Medieval Stimulation

The Medieval era is frequently considered to have ended in the mid 1400s
for various reasons — and all with engineering implications:

(a) *Travel and Multiculturalism*
    Returning Crusaders from the Middle East and the 24-year travels
    of Marco Polo to Asia (1271–1295) provided an alluring broadening
    of geographical and multicultural interests in Europe.
(b) *Trade and Commerce*
    Expanded trade — involving in particular the Silk Road and the
    Hanseatic League — led to considerable commercial activity, also
    providing substantial wealth for some entrepreneurs and emerging
    family dynasties.
(c) *War and Peace*
    The gradual winding down of the 100-Years War in Europe (1337–
    1453) introduced opportunity and resources for the pursuit of
    expanded non-military crafting activities.
(d) *Opportunity and Renewal*
    In the wake of the Black Death, the reduced population throughout
    Europe generated a considerable interest in new and labor saving
    devices.

Further, beginning in the latter period of the Medieval era, a remarkable expansion in various activities had already emerged in support of
the continuing evolution of engineering. Three areas of general activity
in particular need to be noted:

(a) *Agriculture*
    Increasing use of horses and oxen created needs for leather strapping,
    horseshoe fittings, and assorted farming implements such as plows,
    wagons, and various hand tools and farming implements. This cre-
    ated a need for skilled blacksmiths, tanners, wheelwrights, and other
    artisans.
(b) *Machining*
    Expanding use of wood and iron in machined devices required spe-
    cialized tools and knowledge about machine adaptations and their
    maintenance. For example, the previously developed waterwheel was
    increasingly adapted as a useful power source for triphammers and
    bellows, and gravity clocks were subjected to increasing size and
    decorative adornment.

(c) *Seafaring*

The balanced needle boxed compass and windrose appeared in the early 1300s in Europe and provided confidence to seafarers traveling increasing distances off-shore.

The continuing expansion of human initiative, however, had to await the discovery of the movable metal type printing press attributable to Johann Gutenberg (~1454), benefit from the stimulation for oceanic exploration by Henry the Navigator (early 1400), and become enlivened by the rediscovery of the New World by Christopher Columbus (1492), all to be discussed in the subsequent chapter.

4.11 Medieval Engineering: Societal Promotion of Devices

While Ancient Engineering (~8000 BCE → ~500 CE) introduced the important feature of societal recognition of devices, the Medieval period (~500 CE → ~1400 CE) brought about the new feature of an active interest in the promotion of devices by society: cathedral construction was much desired, the domestic and agricultural farming craft trades experienced a substantial promotional interest, the makers of water wheels provided a most effective mechanical power in the service of society, and shipbuilding and maritime commerce became expanded activities. Because of the very important benefits which were perceived to accrue from such device developments, a substantial societal promotion in these human constructions emerged. This supportive promotional activity in the making of ingenious devices represents a positive feedback from society to engineers and can characterize the Medieval Engineering adaptive connectivity by an extension of Eq. (3.11):

$$N(t) \longrightarrow E(t) \longrightarrow D(t) \longrightarrow S(t) . \qquad (4.13)$$

The time line of the evolution of engineering may now similarly be extended, Table 4.1.

Table 4.1   Evolution of engineering now including the Medieval period, ~500 CE → ~1400 CE. We point to another feature of the Content Summary: vertical components suggest considerable integration and the horizontal components suggest forward progressions.

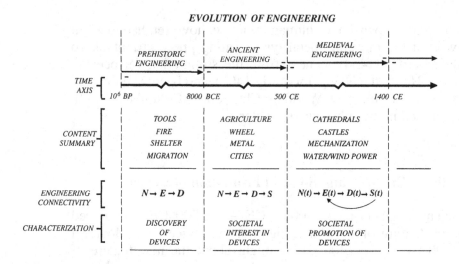

4.12   To Think About

- Medieval cathedrals took centuries to build. Compare and discuss this time-to-design-and-erect with contemporary construction practices.
- Examine the problem of discrete first invention in light of the notion of a continuum of innovation and the extended social contact, including long-distance trade. Choose specific cases.
- It is sometime asserted that the most remarkable feature of Medieval times is the adaptation of power devices as a substitute for human toil and drudgery. Enumerate examples and identify specific features of these devices.

CHAPTER 5

Renascent Engineering

(~1400 → ~1800)

Organizing for Device Production

5.1   Renaissance and Engineering

Beginning about the late 1300s the southern parts of Europe experienced the initial stirrings of a profound movement now known as the Renaissance. This *rebirth* refers specifically to an artistic and intellectual renewal movement which sought to encapsulate the idea that it was time to put aside many of the unenlightened Medieval practices and seek a revival of the more uplifting facets of classical cultures of the Ancient World, especially that of Greece. This idea enjoyed considerable appeal and eventually spread throughout Europe.

For engineering, the Renaissance proved to be highly consequential for three underlying reasons:

(a) *Crafting Proficiency*
    The onset of the Renaissance coincided with the attainment of an accomplished and widespread crafting proficiency.
(b) *Private Wealth*
    Accumulation of private wealth provided capital for architecture, commercial ventures, and patronage of the arts.
(c) *Interest in Novelty*
    Innovative thought stimulated a growing interest in novelty and varied choices of ingenious devices.

Society now also became more inquisitive and adventurous and the ensuing European Industrial Revolution provided considerable opportunity for device development. Indeed, this emerging Renascent Engineering era became the springboard to an Expansive and subsequently Modern age.

5.2   Florence Dome

Like the beginning of the Medieval era with its exceptional Hagia Sophia of ~530 CE, the beginning Renascent period also coincided with a remarkable building project.

In the 1430s, the multitalented Italian craftsman Filipo Brunelleschi became the principal designer and construction supervisor for a most prestigious project of the time: the dome for the cathedral in Florence, Italy. But unlike the dome of Hagia Sophia which was primarily impressive when viewed from its interior, the Florence dome was to be also impressive from afar. This project succeeded most conspicuously for its appealing geometric proportions together with its distant visibility generated considerable public pride and approval — even today. Surviving records indicate that Brunelleschi also introduced some unique and surprisingly modern engineering project control and management methods:

$$
\cdots \rightarrow E(t) \left\{ \begin{array}{l} \textit{project design and planning} \\ \textit{financial and labor force} \\ \quad \textit{management} \\ \textit{activity and material supply} \\ \quad \textit{scheduling} \\ \textit{development of case specific} \\ \quad \textit{tools and techniques} \\ \textit{establishment of an advisory} \\ \quad \textit{oversight committee} \end{array} \right\} \rightarrow D(t)\{dome\} \rightarrow \cdots
$$

$$(5.1)$$

These techniques and practices continue to be used by contemporary engineers.

Engineering and Art

In addition to cathedral dome design and construction, Brunelleschi contributed to a unique technique in art. Until then, paintings displayed objects *as they were* and *not as they appeared to a viewer*: flat and uniform in detail with same size objects in the distance drawn to similar dimensions as those nearby. Based on some geometrical optics considerations of Arabic origin, Brunelleschi demonstrated the use of geometrical perspective — that is the technique of drawing building dimensions in inverse proportions to the distance from the point of view, and with lines of perspective converging to a vanishing point. His architectural drawings had a stunning effect and this technique was thereupon widely adopted.

5.3 Movable-Type Printing

Invention and innovation have always been important to the evolution of changing human perspectives and to the development of new institutions. The trend of Medieval times toward mechanization and iron working, proved to be of particular relevance to the Renaissance because some of the evolving crafting skills provided the essential ingredient for an immediate and socially stimulating interest: movable-type printing. It is as a consequence of this and related development and associated profound societal broadening of thought and practice that engineering became particularly enlivened.

The idea of block printing was first conceived in China about 200 BCE when page-sized stone surfaces were incised with ideographs and then coated with ink for pressing onto paper. In general, these early attempts at block printing were very laborious largely because of the time required for relief carving of complex Asian ideographs; moreover, the images were not particularly clear and prone to fade. In contrast to complex Asian writing, the main alphabet of Europe at the time — the Latin alphabet — possessed fewer than 30 essential symbols for vowels and consonants, 10 numerical characters, and several grammatical and notational signs.

With the advancement of metal crafting, the German artisan Johannes Gutenberg (1400–1468) developed a means of precision casting the mirror images of alphabetic letters at the end of punches made of an alloy of tin, lead, and antimony. These could then be accurately and adjacently mounted — and subsequently reused — on transportable and movable printing plates for placement between the faces of an adapted wine press, Fig. 5.1. And the results were most remarkable: the printed pages were of superb quality and hundreds could be printed in one day.

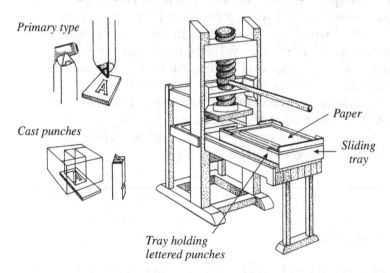

Fig. 5.1   Illustration of Gutenberg type-making and page printing.

By 1454, Gutenberg had organized the first commercial printing operation based on this movable-type concept and other competing printing enterprises soon followed. As it turned out, the demand for written material was phenomenal: only 50 years later about 10,000 printing presses were operational throughout Europe producing in excess of 50,000 titles a year and resulting in perhaps 25 million books in circulation. The demand for printed materials continued to grow and more so with the subsequent print-media innovations of newspapers, periodicals, and pamphlets. Literacy increased from less than perhaps 5% before Gutenberg to some 50% about two generations later. A depiction of this Renascent development is suggested by the connectivity

$$N(t) \begin{Bmatrix} metals \\ fibers \\ wood \end{Bmatrix} \rightarrow E(t) \begin{Bmatrix} design \\ type\ casting \\ typesetting \end{Bmatrix} \rightarrow D(t) \begin{Bmatrix} movable\text{-}type \\ printing\ press \end{Bmatrix} \rightarrow S(t) \begin{Bmatrix} societal \\ fascination \end{Bmatrix}$$

positivefeedback

(5.2)

Presently, nearly 500 million books are printed each year and about 10,000 daily newspapers are produced. Gutenberg's printing initiative thus initiated to a revolution in information dissemination.

It is still common to marvel at the phenomenal impact of the Gutenberg printing device; it proved to be indispensable to the development of consistent spelling, to the emergence of the Protestant Reformation, to the rise of Humanism and Enlightenment, and to the establishment of the Scientific Revolution.

Though the idea of printing a page in one operation had been known as block-printing for centuries, the Gutenberg invention further required the development and integration of several techniques and devices:

(a) Working with soft metals (Lead, Tin, Antimony)
(b) Precision casting of interchangeable alphabetic characters
(c) Development of non-smudging and fast-drying ink
(d) Adaptation of the wine press
(e) Standardization of a multi-stage production process

Until the Renaissance, engineering practice involved a few distinct and empirical design and working considerations. Like the contributions of Brunelleschi, (5.1), the experience of Gutenberg and his movable-type printing press device also changed technical practice, for now a range of skills and knowledge all associated with a device had to converge:

$$
\cdots \to E(t) \begin{cases} \textit{fundraising} \\ \textit{alloy casting} \\ \textit{precision machining} \\ \textit{adaptations} \\ \textit{process definition} \\ \textit{employee training} \\ \textit{marketing} \\ \textit{paper/ink assessment} \\ \vdots \end{cases} \to D(t)\{\textit{printing press}\} \to \cdots
$$

(5.3)

Like the Brunelleschi cathedral dome, the Gutenberg printing press also required in-process changes and continuing problem solving. And so integration of various engineering activities became increasingly important to the continuing evolution of engineering.

Gutenberg's Misfortune

As happens even today, it was financial and managerial shortcomings which proved to be Gutenberg's undoing; less than a year after establishing his printing process — a time when success was apparent — he was removed by his creditors, became marginalized, and died in debt.

5.4   Oceanic Exploration

The earliest prehistoric form of water transport involved logs and rafts of the Sumerians. Soon after, Egyptian reed and wooden boats with triangular or square sails plied the Nile. Expanding Mediterranean trade by the Phoenicians led to merchant ships, and then Persian and Greek naval interests led to multi-tier oared vessels, most also equipped with sails. Next in this progression came the Viking ships providing invasion versatility, and the subsequent Hanseatic Cogs proved to be efficient transporters of goods. With the exception of Phoenician mariners who learned how to navigate on the Mediterranean at night by the moon and stars, most shipping was still limited to coastal sailing.

Towards the end of the Medieval period, European mariners had adopted the magnetic compass — much as it is presently known as a freely rotating balanced magnetized needle with a windrose background in a fully enclosed glass-covered wooden box. They had also adopted the sandglass for short time measurements and developed two additional devices to aid in sailing far from coastal visual reference points: coastal maps and the knotted rope; this rope, with knots tied at agreed upon intervals, had an end-weighted imbalanced attached wooden slab to be laid overboard with knots counted during a specified time interval determined by the sandglass. Ship speed was thus measured in units of *knots* — a speed unit still used by mariners today but now standardized to 1 knot = 1 nautical mile per hour = 1.15 mph = 1.85 km/hr.

Oceanic exploration by Europeans was initially motivated by a very specific commercial interest prompted by crusading knights and other adventurers who had returned from the Middle East with tantalizing objects such as spices, dyes, silks, and porcelain. These goods had their origin in Asia and soon became luxury commodities in Europe stimulating some enterprising Venetian merchants to establish protected trade routes to the Middle East and from there connect to the Silk Road. This commerce became exceedingly lucrative thereby prompting other Europeans to seek their own alternate routes.

Beginning in the early 1400s, Portugal's Prince Henry — also known as *Henry the Navigator* — promoted a national interest in shipbuilding and navigation. Soon the Portuguese *Caravel*, a highly maneuverable ship design with triangular sails and manned by a crew of 20 to 25, began island hopping to the Azores and Madeira Islands, all the while collecting navigational data to be incorporated on maps and tables. Based on this experience, the Portuguese mariners acquired sufficient skill and confidence to explore the west coast of Africa, eventually learning to navigate on the southern hemisphere and discover the Cape of Good Hope. One immediate commercial consequence was the establishment of a marine trading corridor along the western seaboard of Africa as an alternative to camel transport across the forbidding Sahara Desert.

The key to the Portuguese pioneering seafaring accomplishments was evidently their use of the triangular sail on their Caravels. Derived from Egyptian sources, this sail — often called *lateen* — served exceptionally well for tacking against the wind in a zig-zag path thereby allowing near-coast and river estuary exploration; the alternative conventional square sail is very cumbersome for such maneuvers.

About 1450, as a result of Portuguese, Spanish, and Italian shipbuilding and sailing experience, a sturdy ship design emerged with considerable stowing volume and survival capacity against severe storms, thus enabling increasing off-shore and distance sailing. These ships contained three masts with a large central square sail, triangular/combination sails fore and aft, single stern-post rudder, a very sturdy curved hull of butted planking, bow and stern castles, and a crow's nest lookout. This so-called *Carrack* design could be varied in many details and, as it turned out, made history on numerous occasions: Bartholomeu Dias' exploration of the east African coast (1488), Christopher Columbus discovery of the West Indies (1492), John Cabot to Newfoundland (1497), Ferdinand Magellan and global circumnavigation (1519–1522), and many others.

A significant change to the basic Carrack design occurred in the early 1500s when multiple sails per mast, streamlined hulls, and — most importantly — side-mounted cannons were introduced. These so-called *Galleons* served well as both warships and traders and were soon also used on numerous colonization routes by Europeans. Further, trade in African slaves, the transport of Spanish gold and silver from Central and South America, large-scale sea battles, and the simultaneous emergence of the seafaring pirate industry were substantially influenced by the versatility of this exceptional ship design.

Figure 5.2 provides a graphical depiction of ship design changes with time. By analogy to movable-type printing, this process of oceanic exploration may be associated with the progression of material and information flow:

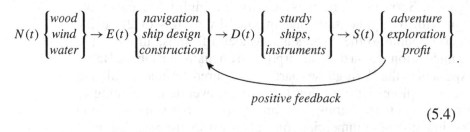

$$N(t) \begin{Bmatrix} wood \\ wind \\ water \end{Bmatrix} \to E(t) \begin{Bmatrix} navigation \\ ship\ design \\ construction \end{Bmatrix} \to D(t) \begin{Bmatrix} sturdy \\ ships, \\ instruments \end{Bmatrix} \to S(t) \begin{Bmatrix} adventure \\ exploration \\ profit \end{Bmatrix}.$$

positive feedback

$$(5.4)$$

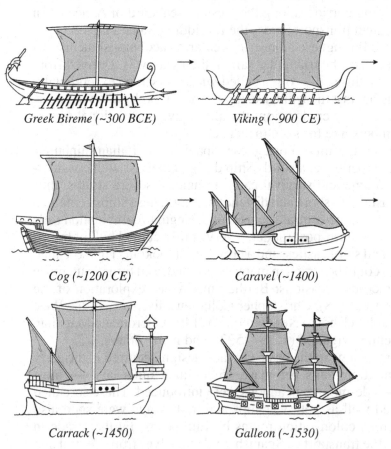

Greek Bireme (~300 BCE)   Viking (~900 CE)

Cog (~1200 CE)   Caravel (~1400)

Carrack (~1450)   Galleon (~1530)

Fig. 5.2   Sailship design evolution of 3000 years (not to scale).

Two Magnetic Crises

Many sailors initially mistrusted the magnetic compass because the needle's autonomous North-South alignment appeared to be like black magic; even some captains chose to use the compass secretly. Then, in the early 1500s, questions about its reliability emerged: in northern parts of the Atlantic ocean, the needle consistently deviated not only from pointing to the North Star but even from the horizontal plane, tending to dip down. What the navigators of the day did not know was the difference between *true north* and the *magnetic north*. This latter unnerving discrepancy initiated the long and arduous inquiry into magnetism, geomagnetism, and its eventual connection with the electric current some 300 years later.

5.5   Intellectual Stimulation

While the making of ingenious devices continued to be most visible and important in this Renascent 1400 → 1800 period, engineering now also became increasingly influenced by the idea of theory, analysis, and experiment to be used specifically in support of practice. Of the many who contributed profoundly to this new perspective during the Renaissance, one can identify several influential individuals with a decidedly mechanistic and analytical viewpoint.

Leonardo da Vinci (1452–1519) was a self-taught Italian polymath who excelled at painting, sculpture, architecture, and engineering. He possessed a particular capacity for the visualization of ingenious interconnected mechanical devices and produced imaginative sketches of bridges, tanks, gear systems, defensive embankments, aeroplanes, and even parachutes and helicopters. Unlike the common practice of the day, da Vinci promoted the use of testing and experimentation as a substitute for guesswork and intuition, and encouraged a methodological process in the design and production of devices. Remarkably, he kept good records and many have survived to the present. His painting Mona Lisa, is perhaps the world's most admired work of art.

A contemporary of da Vinci was Nicolaus Copernicus (1474–1543) of Poland. Following an education in theology, medicine, law, and astronomy, he developed an interest in computational methods of predicting the position of planets which at that time were based on the prevalent geocentric model with the earth taken to be positioned at the center of the universe. Dissatisfied with the existing cumbersome and inaccurate methods, he reintroduced an ancient Greek sun-centered hypotheses and found that calculations using the latest astronomical data yielded much greater accuracy; but this persuasive evidence for an alternative heliocentric characterization of the solar system met with intense opposition from the ecclesiastical establishment because it seemed to decrease the significance of humankind in the cosmos. The Copernican idea that rational thought and free intellectual inquiry — rather than dogma or tradition — should decide on matters of physical reality became a critical stimulus to the ensuing Scientific Revolution.

The Italian Galilei Galileo (1564–1642) possessed an exceptional aptitude for combining theory and experiment. This approach led him to develop aspects of pendulum mechanics, thermometry, gravitational acceleration, structural strength, and both design and manufacture of telescopes together with techniques of observational astronomy. For his defense of the work of Copernicus, Galileo was condemned for heresy by the Church of Rome — a judgment not lifted until the 1990s.

The works of da Vinci, Copernicus, and Galileo, together with an increasing number of singular and creative minds of the Renaissance — Kepler, Descartes, Boyle, Huygens, Hooke, Leeuwenhoek, Newton, Leibnitz, . . . — contributed profoundly to scientific thinking and device development of the Renaissance, Fig. 5.3. An enormous range of intellectual inquiry about natural phenomena and of device development was thus stimulated and soon found to be of increasing significance to the continuing evolution of engineering theory and practice.

Probability and Risk

In the 1600s, a puzzling question became of interest to some recreational gamblers and mathematicians in Europe: how should the accumulated credits in a game of skill and chance be divided if the game had to be terminated before completion? The mystic Blaise Pascal and lawyer Pierre Fermat, both of France, were the brilliant contributors to this area of abstract thought with practical applications, contributing important ideas such as the *degree of certainty* about the future and the role of the ratio of *favorable-to-all-possible-events*. These ideas eventually led to the concept of risk and insurance in the expanding maritime shipping industry, and to the subjects of statistics and probability in mathematics.

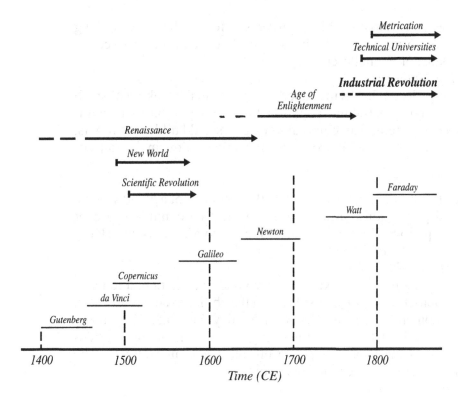

Fig. 5.3 Time line illustration of important contributors and other developments associated with Renascent Engineering.

5.6 Engineering Systematics

Empiricism characterized much of engineering to the end of Medieval times. But then, together with emergent Renaissance thought, engineering practice became increasingly systematized and, concurrently, a complementary theoretical component came to be associated with engineering. This emerging engineering systematic was not sudden but grew progressively and was influenced most notably by the works of the sextet of Brunelleschi, Gutenberg, Prince Henry, da Vinci, Copernicus, and Galileo. While their contributions were distinct, together they formed an integrated whole which is recognizable in the workings of contemporary engineering:

(a) *Filippo Brunelleschi*
    Brunelleschi's records, Sec. 5.2, make it clear that he possessed a thorough understanding and solid organizational grasp of the critical

components of large building projects. Modern builders of long suspension bridges and high-rise buildings are the intellectual descendants of Brunelleschi.

(b) *Johannes Gutenberg*

Gutenberg, Sec. 5.3, was the superb innovator in the making of small devices, paying particular attention to the integration of a multi-component system. Contemporary designers and builders of robots, fuel cells, and process control instrumentation follow in the steps of Gutenberg.

(c) *Henry the Navigator*

Prince Henry, Sec. 5.4, was evidently the visionary of a grand scheme of global proportions, sustaining an unwavering commitment over extended periods of time. The later Guglielmo Marconis and Henry Fords can evidently identify with Prince Henry.

(d) *Leonardo da Vinci*

Here is a person with an exceptionally vivid imagination, a remarkable capacity for graphical display, and the ability to coherently scan an enormous range of knowledge of his day, Sec. 5.5. This capacity for breadth of knowledge and depth of understanding, resides with some of the most productive inventors and innovators — the present-day Renaissance men and women.

(e) *Nicolaus Copernicus*

This humble contributor to the eventual Scientific Revolution, Sec. 5.5, is commonly associated with the notion that rational thought — that is thought based on observation, measurement, analysis of data, tentative formulation of hypotheses, and the testing of these hypotheses — was to be the basis for the theory and practice of both science and engineering. Contemporary engineering has its share of rational thinkers who quietly seek to push the envelope of current thinking and penetratingly question common assumptions.

(f) *Galileo Galilei*

Galileo, Sec. 5.5, was the seminal practitioner of an integrated approach to the study and development of devices — both theory and experiment were necessary. And this dual emphasis on the involvement of both mind and hand has become solidly entrenched in every engineering student's laboratory assignment.

At one level of interest, one might suggest that this sextet of exceptional individuals had independently discovered the cyclical process for solving engineering problems:

$$N(t) \rightarrow E(t) \left\{ \begin{pmatrix} thoughts \\ actions \end{pmatrix} : propose \rightarrow \overset{\frown{modify}}{critique} \rightarrow adopt \right\} \rightarrow D(t) \rightarrow S(t)$$

$$(5.5)$$

At a more subtle level, we suggest an underlying recursive loop incorporating also scientific knowledge as essential to the engineering connectivity progression illustrated in the following:

Recursive engineering thought and actions

$$(5.6a)$$

$$\cdots \longrightarrow E(t) \longrightarrow \cdots$$

and

$$N(t) \longrightarrow E(t) \longrightarrow \cdots$$

$$(5.6b)$$

Nature informing by the scientific method

Adding these recursive and informing concepts to Eq. (4.13) then yields the more complete engineering connectivity characterization to the middle of the Renascent Engineering era, i.e. ~1600:

$$N(t) \longrightarrow E(t) \longrightarrow D(t) \longrightarrow S(t) . \qquad (5.7)$$

The Renascent era was evidently exceedingly diverse and creative; but more changes affecting the theory and practice of engineering were taking place.

5.7 Regional Influences

Some unique initial organizational stirrings of engineering disciplines had already become discernible in the 1400s. At the root of this development was the convergence of sporadic inventions and skilled crafting activities of preceding generations, becoming integrated with ideas on

engineering foundational themes such as strength of materials, importance of testing, and utility of elementary computations. And in this
process of systemization towards a more predictive process of device
development and construction the label *engineer* — as an updating of
the Latin *ingeniator* — slowly became used.

In the early part of the Renaissance, the evolution of engineering
was associated with regional pockets of distinct innovations. For example, Italian architecture and engineering developed an early focus on
building construction characterized by elaborate facades, a skill which
made Italian engineers most desirable by royalty and aristocracy of
other countries. Other engineering initiatives were also pursued so that
by the mid 1400s an extensive compendium on engineering knowledge appeared, containing design details of buildings, bridges, pumps,
siphons, harbors, dams, and surveying instruments. Interestingly, even a
directory of practising Italian engineers appeared.

By the early 1500s, successes of engineering experience were dupli-
cated throughout Europe and various regions promoted particular innovations. For example, the Netherlands — which includes the estuary
of the Rhine and served as a converging trade route for European
crafting products — was increasingly becoming a mercantile banking
center and maritime shipping hub. Here engineers also developed the
systematics of harbor development, river dredging, canal construction,
and land reclamation; additionally, the horizontal axis swivel windmill
was improved and became most effective for pumping water between
canals. Possibly the first higher level instruction on engineering methods and mathematics appeared at the university in Leyden in the early
1600s.

Also at about that time, accessible deposits of copper, gold, iron, lead,
silver, and various minerals were discovered in southern Germany. Mining, together with an increasing interest in ore processing and smelting
became an important engineering focus. By about 1550, a substantial
body of mining experience was recorded, duly published, and soon used
in other mining works in Europe. Some preliminary activity to place
engineering education into a university context also occurred in Giessen,
Germany, in the mid 1600s.

Engineering in France developed an early interest in hydraulics
with water supply projects for cities and palaces, and also canal
construction — including the 300 km *Canal du Midi* which joined the
Atlantic Ocean with the Mediterranean Sea. A unique first coordinated
engineering development occurred with the formal establishment of the
Corps of Military Engineers (1675) and the National School of Bridges

and Roads (1747), both under military authority; the first university-level engineering college was the Polytechnic, inaugurated in Paris in 1794.

Engineering in England was profoundly stimulated by the emergence of the Industrial Revolution, our subject of the next section.

The Calculus

A most powerful mathematical tool for engineers had its beginning in the late 1600s: invention of the differential and integral calculus. The underlying propositions were exceedingly novel when first introduced and consisted of the following:

(a) The physical world could be understood by numbers
(b) Relations between measurable properties of natural phenomena and processes could be specified in algebraic form (e.g. $y = f(x)$, $p = f(V, T)$, etc.)
(c) The Fundamental Theorem of the calculus (i.e. $F'(x) = f(x)$ for $F(x) = \int f(x)dx$) proved to be of great utility

Both Isaac Newton (1642–1727), England, and Gottfried Leibnitz (1646–1716), Germany, worked independently on the so-called *infinitesimal* problem, defined as the limiting tangent to a curve. Though Newton first published his method of *fluxions*, the differential/integral notation now in use was introduced by Leibnitz. Both are considered to be independent creators of the calculus.

5.8 Power of Steam

One of the pivotal innovations in the historical evolution of engineering is the development of devices which relate to the use of steam as motive power.

Heating of water and its consequences had been a source of fascination for centuries. Hero of Alexandria ($\sim$100 CE), a Greek mathematician and inventor, is said to have constructed a rotating sphere with outer tangentially directed tube projections so that when water inside the sphere was heated, the escaping steam from the tubes caused the sphere to rotate. With time, further observations about water and steam occurred and by the mid 1600s, the state of knowledge about water, steam, and the atmosphere could be summarized as follows:

(a) Steam is evaporated water, with a volume increase by a factor of ~1700
(b) When steam cools in a fixed volume, its intrinsic pressure decreases
(c) The atmosphere exerts a constant pressure on all objects
(d) Condensation of steam acts — due to atmospheric pressure — towards volume reduction

Denis Papin, France-England, inventor of the pressure cooker and pressure safety valve, demonstrated in 1679 that a suitably fitted piston could be made to move inside a cylinder by the process of steam condensation. This introduced the remarkable idea of a cylindrical steam engine which, if equipped with appropriately placed valves, might cause a piston to move linearly thereby producing a power stroke for doing work. Papin called this kind of device the Atmospheric Engine.

An important use for a working cylindrical machine was soon identified. This application had its origin in the common national interest for wooden ships, thereby placing a premium on large trees; indeed, Europe had become increasingly deforested of large trees because even a medium size naval vessel required about 2000 mature oak trees; additionally, wood converted into charcoal was a desired fuel in the growing smelting and iron industry. As a consequence, wood had to be used less for burning so that coal mining had become an important activity providing an alternative fuel for applications such as pottery making, iron smelting, farming implements crafting, glass blowing, iron tool manufacture, domestic cooking, and space heating. But there was a severe problem: as in all mining activity, water seepage required pumping water — commonly involving horse-powered continuous chain-of-buckets installation or a slider-crank mechanism attached to a suction-force pump — or mining depth would be severely limited. A powerful cylindrical machine might therefore be developed into an effective water pump. By 1712 Thomas Newcomen of England, produced the first working steam engine and this engine was as impressive as it was simple, Fig. 5.4(a):

(a) Steam from a boiler is admitted into a vertical cylinder equipped with a piston
(b) The steam in the cylinder is then condensed by a jet of cold water
(c) Atmospheric pressure forces the piston downward towards volume reduction of the steam chamber, thereby providing for a power stroke
(d) The cycle is repeated

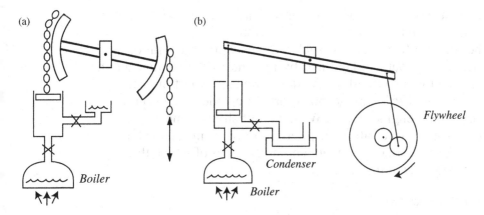

Fig. 5.4   Schematics illustration of the (a) Newcomen and of the (b) Watt steam engines. The symbol $\times$ is to suggest a control valve.

The piston is connected to a cantilever beam so that the piston's downward motion provides for the working stroke for pumping water out of a mine shaft, Fig. 5.4(a). Also, by simple extension, a swivel rod could connect the piston to a wheel thereby producing continuous circular motion as an alternative to the reciprocating rectilinear motion of the cantilever. With steam supplied by a simple coal burning boiler, this steam engine quickly proved to be a suitable power source to pump seepage water from coal mines.

Several important practical features of the Newcomen engine became apparent and proved to be most important:

(a) *Manufacture*
   This steam engine could be built with available materials using common crafting skills.
(b) *Application*
   There existed an immediate and urgent application in the pumping of seepage water from numerous coal mines.
(c) *Cost*
   It appeared to cost less to pump water with such a steam engine — with on-site mined coal as the fuel — than the alternative method of using horses.

Some simple improvements soon led to a convenient characterization. Typically, a steam engine with a 50 cm diameter cylinder and 2 m stroke was able to raise some 50 L of water through a 50 m vertical distance per stroke with a frequency of up to 10 strokes/min.

By the mid 1700s about 100 such engines were in operation in the coal mines throughout England and by the end of the century another 1500 had been manufactured for the domestic and export markets.

Beginning in 1765, about 50 years after the invention of the Newcomen steam engine, James Watt, a Scottish instrument maker and machinist — but most of all an inspired inventor — introduced over the following decades numerous steam engine changes of which three were particularly significant improvements:

(a) *Separate Condenser*
    Steam was transferred by force of flywheel action into a separate condensing chamber thereby allowing the cylinder to operate at a higher temperature, Fig. 5.4(b).
(b) *Double Action*
    A double-acting cylinder with steam entering alternately on both sides of the piston doubled the power-stroke frequency.
(c) *Speed Control*
    A fly-ball governor provided for reliable autonomous mechanical feedback speed control.

Fortunately for Watt, he had early entered into a partnership with a farsighted and successful British industrialist named Mathew Boulton. This provided for steam engine manufacture while Watt devoted himself to its improvements, choosing to restrict himself adamantly to low pressure steam for he dreaded the possibility of a severe boiler explosion.

Watt's inventive contributions increased the efficiency of steam engines from about 2% to 8% and also confirmed this kind of machine as an increasingly reliable and cost effective inanimate source of power, typically available in the range of 1 kW to 200 kW. He also compared the power of his steam engines to that of horses pumping water, thereby introducing the unit of a horsepower. Watt has subsequently been recognized with the SI derived unit of power named after him, 1 Watt (W) = 1 Joule/second; the horsepower (hp) is now defined as 1 hp = 745.7 W.

In a practical sense, it is the works of Papin → Newcomen → Watt which established the primal

$$N(t) \left\{ \begin{array}{c} \textit{water, fire,} \\ \textit{metals} \end{array} \right\} \rightarrow E(t) \left\{ \begin{array}{c} \textit{design, machining,} \\ \textit{assembly} \end{array} \right\} \rightarrow D(t) \left\{ \begin{array}{c} \textit{steam} \\ \textit{engine} \end{array} \right\},$$

$$(5.8)$$

and with it deep and irreversible consequences — not only was the theory and practice of engineering substantially altered but the industrial and social foundations were soon to change as well. And these profound changes were made possible by four basic and connected processes suggested in the following progression:

$$\begin{pmatrix} autonomous \\ energy\ release \\ during\ fuel \\ burning \end{pmatrix} \rightarrow \begin{pmatrix} burning\ fuel \\ converts \\ water\ into \\ steam \end{pmatrix} \rightarrow \begin{pmatrix} steam\ is \\ converted \\ into\ piston \\ motion \end{pmatrix} \rightarrow \begin{pmatrix} piston\ motion \\ is\ converted \\ into\ rotary \\ motion \end{pmatrix}.$$

(5.9)

Such and similar connected conversion processes have become an enduring signature of engineering theory and practice — and all accomplished with ingenious devices.

Evolution of Power Sources

Like the Gutenberg printing press, the steam engine introduced profound changes within society. Moreover, this engine also introduced a new power source readily accessible to serve human interests. Its context relative to other forms of power of engineering interest is suggested by the following progression of power sources:

Solar power
$(\sim 10^{10}\ BP)$ $\longrightarrow$

    Human power
     $(\sim 10^{6}\ BP)$ $\longrightarrow$

        Animal power
         $(\sim 10^{4}\ BP)$ $\longrightarrow$

            Water/wind power
             $(\sim 2000\ BCE)$ $\longrightarrow$

                Steam power
                 $(\sim 1700\ CE)$ $\rightarrow$

(5.10)

The subsequent distinct power source development occurred about 1850 with petroleum sources, followed by nuclear power in the 1950s.

5.9   Industrial Revolution: Technical and Commercial

The emergence of the steam engine introduced a new perspective on human accessibility to inanimate power. Until then, the major controllable inanimate power source was that of water for waterwheels and wind for sails and windmills — even though both possessed important limitations:

(a) Water flow varies seasonally, leading therefore to seasonally varying power production
(b) Wind is subject to diurnal and sporadic changes restricting thereby power production

Additionally, both of these environmentally based energy sources possess evident power density limitations and are site specific, imposing thereby additional limitations. The emergence of the reliable steam engine changed all this for three important reasons:

(a) Delivery of steady power over extended periods of time
(b) Power availability with a range of ratings
(c) Siting flexibility by allowing considerations of distance to markets, proximity of labor, and access to land and sea transportation corridors

These factors provided the power source for the large-scale centralization of the production of consumer and industrial goods, introducing thereby the concept of a factory. Indeed, from $\sim$1750 to $\sim$1800, the number of operating steam engines powering water pumps and crafting shops increased from $\sim$500 to $\sim$2500 in England alone. And in this expansion, engineers proved to be of critical importance, for not only could they build and operate steam engines but they could also design, build, and operate machinery which would be powered by these engines:

$$N(t) \rightarrow E(t) \rightarrow D(t) \begin{Bmatrix} steam\ engine \\ factory\ machinery \end{Bmatrix}. \qquad (5.11)$$

Thus, in these Renascent times, engineers became the builders and operators of a new and more versatile source of power.

We add that many factory developments would continue to use the waterwheel if a suitably flowing river was adjacently available. As happens so very often, a new device may simply provide another option — with acceptance or rejection determined by many factors.

But now there also appeared something unexpected. The power plants and associated production machinery were financed and their products marketed by a new breed of commercial activist: the industrial entrepreneur. This division between the technical builders/operators and financial owners/marketers meant that now engineers became employees. In practical terms, this introduced a trade off between an engineer's assured salary and reduced work autonomy. Thus, while engineers continued to use natural resources and natural phenomena in making devices, it was the employer who specified policy directives related to the technical work of the employee engineers. Using $F(t)$ to represent the generic employer of engineers — that is the firm — we may extend the primal connectivity to read

$$F(t)\{policy\ directives\}$$

$$N(t) \to E(t) \begin{Bmatrix} employee\ technical\ work \\ employee\ assured\ salary \end{Bmatrix} \to D(t). \qquad (5.12)$$

As suggested by the bidirectional arrow between $F(t)$ and $E(t)$, engineers did not thus become passive recipients of the employers policy directives but were expected to provide the firm's directors and managers with appropriate technical information based on their knowledge of $N(t)$ and expected design/manufacturing costs and performance features of $D(t)$. Thus, while a clear distinction between an employers policy directives and employee engineer technical support became established, a remarkable sense of mutualism also emerged. By observation over time, this has proven to be a remarkably stable and enduring arrangement.

Simultaneously, another interesting feature of engineering emerged: the required diversity of engineering skills and knowledge now introduced the modern concept of disciplinary specialization, that is the development of the i-type engineering discipline associated with j-type devices:

$$N(t) \to E_i(t) \to D_{ij}(t). \qquad (5.13)$$

The following suggests such engineering specialities with the approximate time of emergence also suggested.

5.9.1   Machine Engineering (<1650)

Medieval era crafting activities associated with mechanical clocks and iron implements were in place when experimentation and testing with steam occurred. The engineers henceforth provided expanded services in the design and manufacture of cylinders, valves, joints, and flow control instrumentation, together with the specialized machinery for making tools for purpose of manufacture, assembly, and maintenance of mechanical devices. It also spawned increasing precision in other areas, but especially in the armaments industry (e.g. precision boring of steam-engine cylinders is similar to boring cannon barrels). New developments were simultaneously pursued in metallurgy and casting.

5.9.2   Mining Engineering (~1700)

At the time of the first Newcomen steam engine, water seepage limited coal mining to a depth of about 20 m. An effective means of pumping water and the adaptability of the steam engine now permitted deeper and larger mine operations. Hence, attention to rock mechanics, tunnel support, ventilation provisions, and coal transport now became important additional engineering considerations.

5.9.3   Textile Engineering (~1730)

The making of fabric for clothing and sails was a labor intensive activity, originating in Prehistoric spinning and weaving. By early Renascent times, it had become a significant small-batch cottage industry involving, however, considerable variation of fabric quality. Invention of power weaving and spinning — the Flying Shuttle (1733) and the Spinning Jenny (1760) — led to steam-powered factory-based automation thereby revolutionizing the entire textile industry. And with textile mills came the textile engineer.

5.9.4   Structural Engineering (~1770)

Structural materials had not significantly changed since Ancient times: wood, stone, brick, and mortar were the basic building components. But the increasing availability of iron now suggested new construction possibilities to some engineers. To be noted especially is a 60 m bridge across the River Severn at Coalbrookdale, England, assembled from cast iron. Completed in 1779, this remarkable pioneering structure — the forerunner of large iron and steel structures — is still in use.

5.9.5 *Railroad Engineering* (~1800)

Metal rails on which wagons were pulled by horses had already been introduced in England's coal industry about 1790. Placing a steam engine on a wheeled platform with the piston connected to the rim of a wheel seemed self-evident to some inventors. By 1802 the first steam engine locomotive was introduced in an ironworks and by 1825 the first public steam powered railroad became operational in England. The great era of railroad building throughout Europe and in North America was soon underway.

5.9.6 *Marine Engineering* (~1830)

The first modestly successful attempts of steam power for boats began in the 1780s on the Delaware River, USA. To begin, oars were affixed to a horizontal bar and operated by a steam engine with oar motion crudely imitating that of a human rower; but nothing much came of this cumbersome contraption. Robert Fulton of the USA, designed and built the first commercially successful steam powered river boat in 1807; it was 5 m wide, 55 m long, could carry 150 tonnes, was equipped with 5 m diameter side paddle wheels, and operated on the Hudson River between New York City and Albany. More inventors, industrialists, and promoters became involved with river steamboats, also giving rise to the romance and adventure on the Mississippi River. In 1818 the first steam-assisted sail ship crossed the Atlantic, a trip of 28 days.

These various developments in engineering practice occurred even though, with some exceptions and especially that of France, there were no formal and comprehensive engineering schools. Learning on-the-job, a variously practised apprenticeship system, and determined self-inquiry into natural phenomena and machine construction, all combined to serve well in this period of engineering evolution. Initiative, motivation, capacity to learn, and manual skills were the essential ingredients for the young engineer-in-training.

And so, in its formative centuries, engineering moved toward professionalism with one foot firmly planted in the proven empirical crafts with the other foot aggressively exploring the relevance of specialized branches of theoretical and scientific knowledge.

Precision Instrumentation

While industrialization proceeded in several directions, specific interests also emerged in precision instrumentation which addressed particular problems. For example, oceanic transportation suffered considerably because of the absence of a method of longitude coordinate specification; indeed, this deficiency had led to a number of costly and tragic marine disasters. By 1736, John Harrison, England, designed and built the first ship-based spring-powered chronometer of sufficient accuracy and reliability; sea captains now had a means of knowing the time at some reference 0° longitude — typically the Greenwich Meridian in London — which together with conventional optical means of local noon-time specification yielded accurate longitude coordinates. The key relation is that each one hour difference from a reference meridian corresponds also to 15° difference (i.e., 1/24 of 360°) of longitude.

5.10   Industrial Revolution: Social and Demographic

At the beginning of the Industrial Revolution in the mid 1700s, much of Europe could be characterized as predominantly village based with most people living near a subsistence level. With the exception of the iron plow, crop rotation, and use of oxen for plowing, agricultural practices had changed little over the preceding ∼2000 years, but people could point to an increasingly skillful artisan and crafting community — but all based on a small village scale of activity. And into these circumstances there now appeared the powerful steam engine and its profound consequences of centralized factory-based manufacture of a variety of devices generally related to textile, woodworking, metalworking, domestic goods, farming implements, and machine components.

Disruptive consequences of the Industrial Revolution were first felt in England in cottage spinning and weaving as well as in farming. Automation of centralized textile making and large mining operations employing hundreds soon appeared. The associated relocation of agricultural workers and village artisans to the expanded machining and mining operations (exclusively men) and to textile factories (mostly women and children) created severe problems of urban housing and community sanitation, as well as life-style disruptions and family-life interference. Additionally, insufficient municipal services and inadequate labor legislation contributed to appalling conditions both in many places of work and in many homes. As happens so often, this is an example of engineering device developments leading to conditions which outpace a government's ability to anticipate and cope with the process of industry-driven change and urbanization.

To be sure, some concurrent socially positive trends did emerge. Some textile factory owners provided community centers for educational and recreational purposes. The stethoscope, thermometer, and blood pressure measuring devices were introduced. Homecanning, increasing dietary awareness, and more washable cotton as opposed to less washable wool, also became common.

While conditions for many were distressing, some factory owners became very wealthy, *self-made* capitalists. The *laissez-faire* atmosphere did not change until about the mid 1800s when some remedial labor legislation was introduced and municipalities began providing sanitary facilities.

For England, the Industrial Revolution was a phenomenal stimulant, propelling it into a position of global political and economic dominance. By circumstance, it possessed what was required for this kind of revolution to succeed: skilled artisans, efficient administrative infrastructures, good roads and canals, expanding colonial holdings, effective maritime transportation, stable governments, and adequate capital. London grew to a population of about one million by 1800, becoming the largest — and the most polluted — city in Europe. Possessing only 2% of the global population, England soon came to produce about 50% of the world's manufactured goods.

While England became the first commercial beneficiary of the Industrial Revolution, important consequences were pointedly felt elsewhere with the textile industry as one notable example of a global cause-and-effect. The prime mover was the integration of the steam engine, the factory, the Spinning Jenny, and the Flying Shuttle which together enabled England to mass produce textiles of superior quality and at relatively low cost, but now requiring extensive imports of wool and cotton. Many countries then discovered wool and cotton as profitable export products but requiring large farms and cheap plantation labor — and relinquishing their own tactile production capability in favor of low-cost manufactured imports. Indeed, this demand for large tracts of land in warm or tropical regions was so great that it led to large private land holdings, it encouraged the African slave trade, and it entrenched various forms of serfdoms. Further, regions with a ready access to suitable marine harbors now also became connected to a growing global transportation network, commonly involving the flow of raw materials in one direction and finished goods — as well as migrating people — in the other.

Thus, from its earliest inception, the steam engine was the catalyst for some significant changes in the social, commercial, industrial, and geo-political fabric of the world.

The Age of Enlightenment

The latter part of the Renascent Engineering period coincided with some creative innovations in the arts. William Shakespeare (1564–1616), an English playwright, wrote some 40 profound plays and these are still regularly staged. Claudio Monteverdi (1567–1643) of Italy, imaginatively combined drama with music thereby initiating a long chain of operas. Rembrandt Harmensz van Rijn (1606–1669), a Dutch painter, produced portraits of remarkable tonal range and introspection, widely displayed in leading museums and highly valued by collectors. Wolfgang Amadeus Mozart (1756–1791), an Austrian musical genius, produced over 600 major compositions in his short life and these are still regularly performed in the world's concert halls. Throughout Europe, intellectuals began meeting in cafes to discuss artistic, social, and political issues. And in the British Colonies of North America, people met to discuss political independence from England — declared in 1776.

5.11   Renascent Engineering: Organizing for Device Production

The Renascent period ($\sim$1400 $\rightarrow$ $\sim$1800) emphasized some important considerations to the theory and practice of Medieval Engineering, both of which enhanced the production of devices. The first was the extended and systematic incorporation of recursion in device design and development, $Re(t)$, as an addition to Eq. (4.13):

$$Re(t)$$

$$N(t) \longrightarrow E(t) \longrightarrow D(t) \longrightarrow S(t). \tag{5.14}$$

Second, the emergence of science and the scientific method became a more focussed part to the engineering connectivity network

$$N(t) \longrightarrow E(t) \longrightarrow D(t) \longrightarrow S(t), \tag{5.15}$$

$$Sc(t)$$

with $Sc(t)$ representing the feedforward scientific method of *nature informing* the theory and practice of engineering. And third, the function of the engineer as an employee emerged

$$F(t)$$

$$N(t) \longrightarrow E(t) \longrightarrow D(t) \longrightarrow S(t), \tag{5.16}$$

with $F(t)$ representing the generic firm as the employer of engineers.

Additionally, the first adverse environmental impact began to be recognized during the Industrial Revolution part of the Renascent engineering period: the increased burning of coal in support of the growing smelting, refining, and steam-power manufacturing industry generated a corresponding increase in combustion gases and airborne particulate products. At that time, little thought was given to sequestering the effluents so that air, water, and land became locally contaminated. This adverse external impact can be incorporated into the engineering progression by a loop emanating from the resource extraction linkage:

$$\textit{Environmetal Contamination}$$

$$N(t) \longrightarrow E(t) \longrightarrow D(t) \longrightarrow S(t). \tag{5.17}$$

Note that environmental contamination constitutes an alteration of nature $N(t)$.

Table 5.1 Addition of Renascent Engineering ($\sim$1400 $\rightarrow$ $\sim$1800) to the time line of engineering evolution. The content summary now forms a 4 $\times$ 4 matrix of interrelated components.

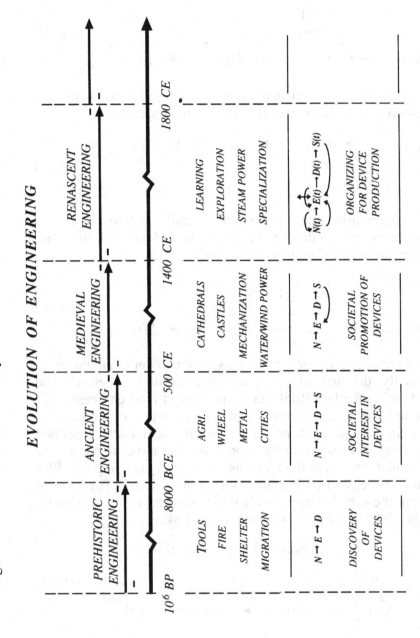

EVOLUTION OF ENGINEERING

	PREHISTORIC ENGINEERING	ANCIENT ENGINEERING	MEDIEVAL ENGINEERING	RENASCENT ENGINEERING
	10^6 BP	8000 BCE	500 CE	1400 CE	1800 CE
	TOOLS	AGRI.	CATHEDRALS	LEARNING
	FIRE	WHEEL	CASTLES	EXPLORATION
	SHELTER	METAL	MECHANIZATION	STEAM POWER
	MIGRATION	CITIES	WATER/WIND POWER	SPECIALIZATION
	$N \rightarrow E \rightarrow D$	$N \rightarrow E \rightarrow D \rightarrow S$	$N \rightarrow E \rightarrow D \rightarrow S$	$N(t) \rightarrow E(t) \rightarrow D(t) \rightarrow S(t)$
	DISCOVERY OF DEVICES	SOCIETAL INTEREST IN DEVICES	SOCIETAL PROMOTION OF DEVICES	ORGANIZING FOR DEVICE PRODUCTION

Collectively then, Renascent Engineering can now be represented by the engineering connectivity with straight arrows representing material and energy flow and curved arrows representing information and ingenuity flow,

$$N(t) \longrightarrow E(t) \longrightarrow D(t) \longrightarrow S(t) , \qquad (5.18)$$

thus yielding a comprehensive extension to Eq. (4.13). Thus, the profession of engineering became increasingly multifaceted.

Table 5.1 incorporates these evolutionary aspects of the emergence of engineering and thus characterizes Renascent Engineering, $\sim 1400 \rightarrow \sim 1800$.

5.12 To Think About

- Brunelleschi and Gutenberg may, in a particular sense, be considered to be the first modern well-rounded engineers capable of managing important engineering aspects of a technical project. Discuss, compare and relate their works to contemporary engineering practice.
- How did the Industrial Revolution contribute to such major social-political-economic developments as Colonialism, Capitalism, and Socialism? Be specific.
- Renaissance scientists and engineers began defining energy and power as both a physical phenomena and as an analytical concept. Recognizing an integral/differential connection between power and energy, determine circumstances in engineering which makes one of these terms the more relevant parameter.

CHAPTER 6

Expansive Engineering

(~1800 → ~1940)

Environmental Impact of Devices

6.1 Progress and Engineering

The 1800s were remarkable in that Western society adopted a most notable sense of technological optimism. To be sure, part of this optimism can be traced to the ancient idea of *progress* which sought to project the general ideal of a forward movement of society towards a more preferred future. An influential utilitarian perspective to this notion had already been introduced by the British scholar Francis Bacon (1561–1626) who promoted the notion that knowledge about the physical world could be used for the betterment of the human condition. But the remarkable feature of this 19th century optimism was the enmeshing of a broad societal interest in aspects of engineering and in a new physical-biological characterization of reality. Some particular aspects of this broader perspective are readily identified:

(a) *Transportation*
   By the early 1800s, the steam railroad and soon after the steam ship, provided a more extensive and reliable means in support of commerce, trade, and migration. Indeed, a dynamic transportation *web* began encircling the globe.

(b) *Urbanization*
   A profound urbanization process emerged in the mid-1800s, creating unprecedented demands and opportunity for city housing, urban transportation, and sanitary installations. Some countries even became ~50% *urbanized*.

(c) *Science and Medicine*

By the end of the 1800s, molecular and atomic structure — with a faint hint of nuclear theory — became recognized and the far-reaching bacterial basis of illness and epidemics became convincingly established. New *worlds* were opening for the deeply inquisitive minds of the day.

(d) *Social Action*

An increasing interest in coordinated public action also characterized the 1800s: some absolute dynasties were toppled, representative governments emerged, labor unions were organized, Marxism appeared, new religious denominations arose, women suffrage had its beginnings, and the movement to abolish slavery gained in strength. Individuals became *empowered*.

Superimposed on these developments was the increasing realization that — for the first time in human history — individuals were not born into a rigid and confining social class totally at the mercy of the few and powerful, but that initiative, learning, and hard work, could actually allow the shaping of personal futures. An enormous *awakening* was taking place.

Engineers of the day could well direct their skills and aspirations to the expanded making of especially ingenious devices and thereby encouraging various progressive notions. Indeed, if a first event could be identified as encapsulating a most visible form of this technological optimism, then the 1851 Great Exhibition of Industrial Production may be considered as the signalling and widely noted stimulant. This first ever international industrial fair — indeed the first World Fair — was held in London in an exhibition hall of revolutionary design: a vast glass and iron pre-fabricated building soon labelled the Crystal Palace. It was in this airy space that European manufacturers proudly displayed the latest steam engines, machine tools, instrument dials, valves, pipes, control devices, pulleys, pumps, gears, navigational instruments, agricultural implements, mining equipment, mill machinery, hand tools, household goods, electrical devices, testing equipment, woodworking machines, optical lenses, ..., and the list went on. The world had never before seen such a collection of human inventiveness, and all displayed in a lofty and expansive building. Progress now became associated with a diversity of devices emanating from factories and all promising some direct beneficial service to the individual and community.

In these emerging days of expansive thinking and doing, engineers were a proud and confident community. Not only was the term engineering increasingly recognized and esteemed, but engineers were willing to produce myriad devices which society was eager to acquire. Indeed, some engineers now also became active in the promotion of their works including participation in fledgling trade and technical associations. This Expansive 1800 → 1940 period could well be renamed the *exponential growth* engineering era — and with it came the increasing recognition of engineering as a vital profession.

The Bicycle

Though da Vinci had already conceived of a bicycle, practical versions did not appear for another 300 years, beginning in the 1820s as a two-wheeled walking device with a seat and swivelling steering mechanism, to be successively improved with pedals, chain, and gears. The essential features of a modern bicycle first appeared in 1885. World wide production of bicycles has increased steadily to a current rate of ∼100 M/year with China alone producing about 40 M/year.

6.2   Steam Railroad

The original use of the steam engine, developed and continuously improved during the 1700s, was to pump water out of coal mines. It then became the dominant means to power factories. Soon it had become evident that a steam engine on a platform equipped with wheels could become self-propelling. Indeed, already in 1769 a French steam-powered 3-wheeled gun carriage became barely operational and in 1825 the first public railway entered service near London. By 1840, a locomotive boiler design appeared with multiple horizontal tubes providing a larger heat transfer area for maximum steam production and hence reduced fuel consumption. This basic design, subject to numerous component modifications over time, remains operational in some countries even today. Figure 6.1 provides a graphic depiction of the evolution of early steam transportation.

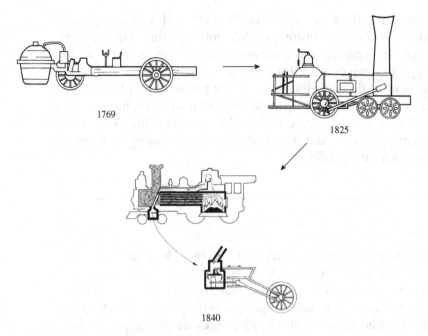

1769

1825

1840

Fig. 6.1 The first working steam locomotive was designed as a three-wheeled gun-carriage for road use, 1769. By ~1825 a number of locomotive designs were manu-factured. A horizontal tube, swivelling-trunk, and with various wheel combinations, became the common design beginning about 1840, lasting about 100 years.

By about 1820, the concept and expected utility of steam railroad-ing over longer distances had become widely recognized. Governments were ready for appropriate land-use legislation and also provided funding support. With considerable haste, right-of-ways were surveyed, railway bridges were built, tunnels were bored, tracks were laid, and stations built. By 1835, some 4000 km of track was in place in Europe and 2000 km in North America. In 1869, the east and west coasts of the United States were linked by railway, the Trans-Canada Railway was completed in 1895, and the Trans-Siberian Railway entered service in 1904. National interests in unification and regional economic consequences became associated with these steam railroad projects, well represented by the primal

$$N(t) \begin{Bmatrix} water, \\ iron, \\ coal, \\ land \end{Bmatrix} \rightarrow E(t) \begin{Bmatrix} design\ and \\ manufacture, \\ transportation \\ networks \end{Bmatrix} \rightarrow D(t) \begin{Bmatrix} steam \\ powered \\ locomotives, \\ trains \end{Bmatrix}$$

(6.1)

also providing much opportunity for creativity and ingenuity.

The Iron Horse

Railroading of the mid 1800s to early 1900s captured the imagination of many writers and artists. Poets, novelists, and journalists glowingly wrote about the heroic and powerful *Iron Horse* and many paintings featured a puffing and gleaming locomotive in a picturesque mountain setting. For isolated towns, the railroad provided assurance of survival and economic importance; for the heavy machine industry and railroad tycoons, it provided wealth and influence. Eventually, the introduction of Pullman Cars and the Orient Express added luxury and mystique to the railroad. Even railroading language formed part of Western culture: it became important to *get on track*, sometimes to *let off steam*, always to maintain one's *train of thought*, but never to *fall asleep at the switch*.

6.3 Materials Processing

But all was not well. The railway network was growing more rapidly than justified by the available iron technology. Railway tracks were failing, steam pressure vessels were bursting, and a general recognition emerged that it was the quality of iron which was unsatisfactory. Indeed, the armaments industry had for the preceding 400 years been hampered by dangerously failing cannon barrels and had long been interested in improved metals.

Materials widely used by the early 1800s — wood, stone, adobe, fiber, leather, bronze, glass, iron — had already been pioneered in Prehistoric and Ancient times and had changed little during the intervening millennia. With respect to iron, all one could find were parts of the countryside littered with belching foundries and sooty smelters, all producing iron of uneven quality and in small batches. And this did not meet the demand of the times.

But gradually the metallurgy of iron and steel became clearer. For example, it was established that the commonly produced brittle cast iron contained a $\sim 4\%$ carbon-to-iron ratio but a $\sim 0.6\%$ ratio with controlled trace elements of rare metals yielded a much harder and more durable steel.

6.3.1 *Hard Steel*

In 1856, Henry Bessemer of England patented a method for the large-scale production of high quality steel. The inventive idea was to force air through molten iron so that the carbon in the iron could combine with the oxygen from the air to form gases CO and CO_2, which would then exit from the molten material. Further, the carbon–oxygen reaction in the melt is exoergic, thereby leading to higher process temperature. By careful temperature control and including the selective addition of some rare metals, high grade low carbon steel could be tailor made for particular applications. This was a major breakthrough yielding high quality steel in large quantities and rapid production rates — and all at relatively low cost. The corresponding primal is suggested by

$$N(t) \left\{ \begin{array}{l} iron, \\ oxygen, \\ heat \end{array} \right\} \rightarrow E(t) \left\{ \begin{array}{c} chemical\ reactions, \\ atom/molecule \\ migration, \\ high\text{-}temperature \\ control \end{array} \right\} \rightarrow D(t) \left\{ \begin{array}{c} high\ grade \\ steel \\ products \end{array} \right\}$$

$$(6.2)$$

to illustrate a dominant focus of Materials Engineering beginning in the mid-1800s.

6.3.2 *Synthetic Dye*

In addition to metallurgical processing, chemical processing became of considerable relevance about the same time, primarily affecting the textile industry. Textiles involved vegetable fibers (flax and cotton) and animal fibers (wool and silk), were often colored using a small selection of available natural plant-grown and marine-life dyes; but these dyes were limited, prone to fade, and their colors were often difficult to reproduce. In 1856 and after, William Perkins, England, discovered that the selective distillation — and further redistillation — of coal tar or other selected petroleum grades could yield deep violets, bright reds, shining blues, and clear green dyes, any of which would bind to fiber and thus be long-lasting. Further, these synthetic dyes could be inexpensively produced thereby contributing to significant growth of the textile industry. Here, too, one may identify the underlying primal,

$$N(t) \left\{ \begin{array}{c} fibers, \\ hydrocarbons \end{array} \right\} \rightarrow E(t) \left\{ \begin{array}{c} distillation, \\ mixing \end{array} \right\} \rightarrow D(t) \left\{ \begin{array}{c} colored \\ textile \end{array} \right\},$$

$$(6.3)$$

forever changing textiles and fashions, and giving birth to the organic chemical industry.

6.3.3 *Vulcanized Rubber*

Also in the mid-1800s, a chemical process of treating naturally available rubber was investigated and was to have a most pronounced impact on the future of transportation. To begin, it was known that natural rubber, consisting of air-cured sap which oozes from the bark of the South American tree *Havea brasiliensis*, became brittle at moderately low temperature and sticky at moderately high temperatures. In 1830, a determined American experimentalist and hardware merchant discovered that mixing this natural rubber with sulfur and heating the resultant mixture, yielded a stable compound which remained elastic over a larger temperature range. The inventor, Charles Goodyear, called this process *vulcanization*. Unknown to Goodyear, he had discovered a way of producing a synthesized elastic polymer. For this invention, one may evidently write

$$N(t) \left\{ \begin{matrix} natural \\ rubber, sulfur \end{matrix} \right\} \rightarrow E(t) \left\{ \begin{matrix} mixing, \\ heating \end{matrix} \right\} \rightarrow D(t) \left\{ \begin{matrix} synthetic \\ rubber \end{matrix} \right\} \quad (6.4)$$

as the ingenious primal.

6.3.4 *Dynamite*

Another significant chemical process of the mid-1800s was the taming of nitroglycerine — an unstable and dangerous explosive compound obtained by mixing glycerol with nitric and sulfuric acids. Alfred Nobel of Sweden, made the chance discovery in 1866 that mixing nitroglycerine with a compound called *kieselguhr* — an abrasive siliceous natural substance now found useful in battery insulation — yielded a very stable mixture. Such a combination, which Nobel called Dynamite, could more safely be used as a powerful explosive but the mixture now required a detonating device. This innovation constitutes the primal progression

$$N(t) \left\{ \begin{matrix} various \\ materials \end{matrix} \right\} \rightarrow E(t) \left\{ \begin{matrix} mixing, \\ trigger \end{matrix} \right\} \rightarrow D(t) \left\{ \begin{matrix} controlled \\ explosive \end{matrix} \right\}, \quad (6.5)$$

indicating also the importance of compensatory device components.

6.3.5 *Petroleum Products*

Finally, significant quantities of petroleum and selected derivatives first emerged in the mid 19th century (1857, ON, Canada; 1859, PA, USA). Then in the late 1800s, natural gas was discovered in the USA, providing a more suitable substitute for coal gas. More importantly, larger oil fields were soon discovered (1901, TX, USA; 1908, Persia, Arabia, ...) prompting large investments in the technology of oil extraction, oil transportation, and fractional distillation to yield various product streams for particular purposes. Three of the early products were kerosene for lamps, moderate octane gasoline suitable for the emerging internal combustion engine, and oil for lubrication purposes. The primal

$$N(t) \begin{Bmatrix} gas, \\ liquid \end{Bmatrix} \rightarrow E(t) \begin{Bmatrix} exploration,\ extraction \\ distillation \end{Bmatrix} \rightarrow D(t) \begin{Bmatrix} fuels, \\ lubricants \end{Bmatrix}$$

$$(6.6)$$

proved to be most far-reaching. Indeed, the consequent triumvirate industries of (a) gasoline supply, (b) automobile manufacture, and (c) road construction, soon established themselves as a mutually stimulating and symbiotic enterprise. They jointly expanded in the 20th century, eventually resulting in the very extensive gasoline distribution, automotive manufacturing, and highway network in Europe and North America, and subsequently in much of the remaining world.

Engineering Stimulating Science

The above several examples of 19th century materials processing came about as a result of inspired intuition, chance observation, and systematic testing. They had the interesting effect of stimulating the sciences of inorganic chemistry and molecular physics. Recall also that the steam engine of the 18th century stimulated the science of thermodynamics. Thus, while the birth of engineering preceded that of science by millennia — $\sim 10^6$ BP versus mid-1500 — science and engineering have subsequently established a most effective symbiosis.

6.4 Steam Shipping

The entry of steam power into marine shipping appears to have proceeded in a more measured way than that of steam power for transport on land. For many centuries, wood for hulls and wind for sails were the essential requirements for oceanic transport though wind damage and wind irregularity had long been recognized as severe risks. Steam power for river transport seemed like a more immediate consideration and by the early 1800s steam powered flat-bottomed paddle wheelers were routinely operating on some rivers, especially on the Mississippi River and the Hudson River in the USA. Ocean vessels, however, soon followed with the first steam-assisted sail ship crossing the Atlantic in 1818.

While the trend toward steamships had begun in the early 1800s, captains and shipbuilders began to realize some distinct advantages associated with iron hulls as opposed to wooden hulls:

(a) Iron hulled ships were less likely to break-up upon grounding.
(b) Riveted iron hull plates avoided the perennial structural and leakage problem of end-to-end wooden plank joints.
(c) Deforestation throughout Europe and subsequently along the eastern seaboard of North America rendered suitable wood increasingly scarce and hence costly.

And so a gradual shift to both iron-hulled and steam-driven paddlewheelers emerged though for ocean-going ships, masts and sails were also initially installed and used. The first fully steam powered Atlantic crossing occurred in 1838, a trip of 18 days — typically half the time of fast sailships under favorable wind conditions.

Paddlewheels versus Screw Propellers

The pioneering steamships of the early 1800s were equipped with side- or stern-mounted paddlewheels since craftsmen had some 2000 years experience with waterwheels. However, during rough seas these paddlewheels provided uneven propulsion and even from efficiency considerations there were questions. For that reason numerous propeller designs emerged and some were installed on small boats. A decisive test for large ships occurred in 1845 when the British navy subjected a sidewheel warship and a propeller warship, each equipped with a comparable steam engine, to a tug-of-war. Propeller propulsion won by a decisive margin.

By the end of the 1800s, oceanic commercial and military sail ships had convincingly yielded to steam powered propeller driven ships. This development conforms to the primal

$$N(t)\left\{iron,\ coal\right\} \rightarrow E(t)\left\{\begin{array}{l} marine\ design,\\ rotational\ power \end{array}\right\} \rightarrow D(t)\left\{steam\ ships\right\},$$

(6.7)

with bunker petroleum eventually replacing coal as the more convenient fuel.

6.5 Cardinal Transitions

It occurs now and then that some device developments are especially consequential because they identify multi-purpose transitions leading to a vastly diversified opportunity for engineering thought and actions and, in a subtle but very consequential way, may have a profound impact on society. We call these *cardinal transitions* and describe three such cases identified with the Expansive Engineering era, $1800 \rightarrow 1940$. The specific cases chosen tend to emphasize features of artifact, method, and process.

6.5.1 *Cardinal Artifact: The Turbine*

Recall that the original steam engine involved a reciprocating piston to induce pulsating water flow in draining mines. Its use in locomotion and shipping then required the conversion of reciprocating motion of the piston into rotational motion of traction wheels or paddle wheels, or propellers for linear motion of a ship; in each case, a flywheel function smoothed out the pulsing power stroke of the piston. Two considerations now became evident: efficiency improvements could be expected if the energy losses due to alternating acceleration/deceleration of the reciprocating piston could be eliminated and repeated oscillatory motion tends to limit crankshaft rotational speed thus affecting power output.

Interestingly, using linear motion of a jet of steam to induce rotary motion in a sphere had already been explored — and possibly even demonstrated — by Hero of Alexandria nearly 2000 years ago. Then, in 1629, Giovanni Branca, Italy, described how a jet of steam impinging on vanes mounted on the rim of a pivoting metallic disc could induce high speed rotational motion directly. And in 1884, Charles Parsons, England, demonstrated a multi-jet steam turbine based on the Branca concept rotating at the then phenomenal speed of 18,000 rpm. Its emerging application was to rotate dynamos for electric current generation which, by 1900, powered community electricity generators. Improvements eventually yielded efficiencies of ~40% which may be compared to ~10% for piston-based steam engines. And, of surprising interest, the steam turbine possessed only one moving part.

A further capability of the turbine concept occurred some 40 years later in its adaptation to jet powered aircraft. As shown in Fig. 6.2, the turbine proved to be a very *fertile* device for industry:

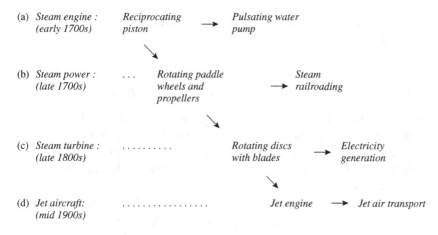

Fig. 6.2   Connectivity depiction of the turbine.

Steam Turbine Powered Ships

About 1895, Parsons adapted a gear box to his high speed 1.5 MW steam turbine to power a 40 metric ton ship capable of higher speed than conventional vessels. He provided a convincing demonstration of the effectiveness of this new device at a navy sail-past attended by Queen Victoria: quite uninvited, he intruded and then outraced the fastest navy destroyer 35 to 27 knots; the Queen was *not amused* and the navy brass were *fuming* — but the public was entertained. Soon, however, new warships were designed to be powered by steam turbines.

The steam turbine and subsequent gas turbine constitutes a most remarkable engineering development. Though few people have actually seen the inside of such a device, it generates most of electrical AC power. It also represents an exceptional culmination of the historical development of power sources:

$$\begin{pmatrix} \textit{Wind/Water Rotation:} \\ \textit{power: } {\sim}100\,W\,to\,{\sim}1\,kW \\ \textit{efficiency: } {\sim}10\%\,to\,{\sim}30\% \end{pmatrix} \rightarrow \begin{pmatrix} \textit{Steam Engine:} \\ \textit{power: } {\sim}1\,kW\,to\,{\sim}\,1\,MW \\ \textit{efficiency: } {\sim}2\%\,to\,{\sim}15\% \end{pmatrix} \rightarrow \begin{pmatrix} \textit{Steam/Gas Turbine:} \\ \textit{power: } {\sim}1\,MW\,to\,{\sim}1\,GW \\ \textit{efficiency: } {\sim}30\%\,to\,{\sim}50\% \end{pmatrix}.$$

$$(6.8)$$

Note also the implied variations of mobility and transportability of these three power device groupings.

6.5.2   Cardinal Method: Interchangeable Parts Manufacture

Transitions are not limited to devices; changing methods may also be most consequential. One such major transition occurred in America during the mid-1800s and was subsequently called the American System of Manufacture. Some prior comparative conditions on both sides of the Atlantic need to be noted.

By the late 1700s, various European regions had established a variety of crafting guilds. The members of these guilds had, through years of apprenticeship and journeyman experience, learned how to begin with suitable raw materials, fashion rough component parts, and then by skillful shaping and fitting produce individualized devices. The small but unique differences in detail of these hand-made objects reflected on the skill of these artisans and served to institutionalize certain trades and labor practices. Intense resistance by guild members to changes in the way devices were produced existed because such changes threatened the relevance of years of training and experience.

In North America, conditions and practices were very different: much land, ample resources, scarcity of skilled labor, and an open trade-labor market mostly free of ingrained traditional ways of producing devices. Indeed, by the early 1800s house construction was of the balloon-frame type avoiding thereby time consuming European-style mortise-and-tenon fittings, interior walls were boards rather than plaster, and furniture was suitably log-based and appropriately called Early American style. Arable land development, lumbering, millwork, smithing, iron smelting, and wheel-and-wagon making were still the important labor intensive activities. Pragmatism and ingenious adaptation were the guiding principles.

For the few water and steam powered factories and machine-based industries in the New World — located primarily near the north-eastern seaboard of the USA — the labor situation was more acute. For whatever skilled machinists or machine operators could be found at the immigration docks, they were quickly shunted to the spinning and weaving mills and to the assorted iron and implements industry. By any standards, the ratio of skilled-to-unskilled workers was most unfavorable.

But then, a distinct innovation in the manufacture of devices began to be accepted. It consisted of two considerations:

(a) Might not the few available skilled craftsmen be used to produce sufficiently precise machinery to be operated by the unskilled?
(b) Then, might not precision standardized parts so produced in quantity be then assembled into more complex devices with the work again done by the unskilled?

Thus, the idea of interchangeable parts manufacture became established.

An early effort to produce sufficiently accurate interchangeable parts was funded by the USA Federal Government for purpose of making small arms. This initiative had a bumpy start but eventually succeeded. Indeed, it succeeded sufficiently to persuade private industry to adopt this method of interchangeable parts manufacture. Devices such as wooden clocks, farm harvesters and reapers, bicycles, sewing machines, tools, implements, typewriters, etc. were produced in quantities by this method. By the late 1800s, the American engineer Frederick Taylor introduced the concept of time-motion study in order to improve efficiency of this new form of production.

Though European guilds looked upon this method of making devices with much disdain, interest in this method of interchangeable parts manufacture soon emerged. International commissions came to observe and manufacturing experts came to study. Several countries even purchased such machines to establish their own American System of Manufacture.

A most remarkable large scale demonstration of this new way of manufacture was realized by Henry Ford for low-cost high-volume automobile production in the early 1900s. Interchangeable parts manufacture, moving conveyor assembly, and mass production proved to be the ingredient for a unique American innovation.

The Trajectory of New Methods

While the USA became the great practitioner of interchangeable parts manufacture, this method was actually born in Europe though not widely implemented. A useful reference point may be taken to be the early 1800s when the British locksmith Henry Moudslay developed some precision woodworking machinery readily adaptable to the making of marine block-and-tackle. Large numbers of such pulleys were required, ~2000 per ship, and hand manufacture was tedious and time consuming. But four of Moudslay's machines, operable by unskilled labor, could produce over 100,000 per year. A point to note is that methods of how to make devices may cross national boundaries more readily than devices themselves.

6.5.3   *Cardinal Process: Synthetic Fertilizer Production*

Nitrogen belongs to the set of absolutely essential elements of healthy plant and animal life: oxygen, hydrogen, carbon, nitrogen. It enters into living objects by transit from fertile soil and water into plant roots by bacterial catalysis, always bound in some molecular form. Humans digest these edible plants, with nitrogen then appearing in amino acids and subsequently form critical components of proteins. Finally, the proteins enter the human cell with the nitrogen becoming essential to the formation of DNA.

Only 0.02% of the earth's crust consists of nitrogen though in arable soil it may be supplemented by compost, manure, and the recycling of leguminous plants such as peas, soybeans, peanuts, and potatoes. Hence, the nitrogen initiated heterogeneous progression of critical biological relevance

$$\begin{matrix} nitrogen \\ in\ soil \end{matrix} \rightarrow \begin{matrix} nitrogen \\ in\ plants \end{matrix} \rightarrow \begin{matrix} human \\ digestion \end{matrix} \rightarrow \begin{matrix} amino \\ acids \end{matrix} \rightarrow protein \rightarrow DNA$$

$$(6.9)$$

is evidently bounded in quantity by the very limited nitrogen naturally available in the soil.

By the late 1800s, a profound and most worrisome realization emerged: in light of the growing world population — and it had increased from ~1 billion to ~2 billion in that century — the total naturally accessible nitrogen via Eq. (6.9) was recognized as inadequate in the intermediate terms. Indeed, already at that time, some European countries had virtually used all their arable land and still had to import wheat. With the global population showing no signs of declining, a nitrogen deficiency in the human diet thus suggested the real possibility of malnutrition, stunted human development, and premature death by starvation — and occurring in the foreseeable future.

At that time in the late 1800s, it was known that 78% of the atmosphere consisted of inert diatomic nitrogen, N_2. Further, it was also known that synthetic fertilizer consisting of varying proportions of potassium, phosphorus, and nitrogen, could increase agricultural yield by a factor of up to four. With potassium and phosphorous adequately available from terrestrial sources, the extraction of nitrogen from the atmosphere now became a most acute research activity.

Over the period 1906 to 1909, Fritz Haber of Germany conducted experiments ultimately resulting in the laboratory extraction of nitrogen from the atmosphere and having it appear in the loosely bound molecule ammonia, NH_3. The device consisted of a cylindrical converter with an inflow of air and hydrogen passing over a suitable metallic rare-earth catalyst, under modest temperatures of $>500°C$ but high pressure >50 MPa (i.e. 500 atm) ultimately yielding NH_3 by the catalysis

$$N_2 + 3H_2 \rightarrow \left(\begin{array}{c} catalytic \\ process \end{array}\right) \rightarrow 2NH_3 + (\ ). \qquad (6.10)$$

This device was scaled-up for industrial purposes by fellow countryman Karl Bosch with the process now known as the Haber-Bosch Synthesis; Haber was awarded the Nobel Prize of 1918 for this discovery of nitrogen extraction from air by means of catalytic conversion.

Both WWI and WWII interfered with the orderly development and application of synthetic fertilizer; indeed, most nitrogen so produced was initially used for the production of explosives. But by the 1950s, atmospheric nitrogen became increasingly available as the soluble crystalline compound urea, $CO(NH_2)_2$, and primarily manufactured for agricultural fertilizer purposes; lesser quantities were used stark as industrial and pharmaceutical feedstock. Presently, some 10^9 kg/year of nitrogen-based fertilizer is produced globally, with most countries having their own plants based on the Haber-Bosch process.

Significant Inventions

The most profound consequence of this process of nitrogen-based fertilizer manufacture is the increasing recognition that without it, about 1/3 of the present world population would not be alive. In such terms, the Haber-Bosch catalytic converter probably qualifies as the most consequential invention, ever!

6.6   Automotive Transport

The steam engine of Papin–Newcomen–Watt is an *external combustion* engine; that is, an exoergic chemical reaction occurs outside the working cylinder. However, in 1673, Christiaan Huygens of the Netherlands first demonstrated the concept of *internal combustion* by exploding a small charge of gunpowder inside a metal cylinder and thereby moving a piston; this idea proved to be impractical at the time but about 200 years later a connection was made with three technical developments:

(a) Availability of combustible coal gas and processed petroleum fuels
(b) Battery-supplied electrical currents to provide sparking
(c) Practicality of fuel-air mixture ignition by an electrical spark

The feasibility of using these innovations to improve on the Huygens internal combustion concept was demonstrated in the late 1800s — and soon led to some most important consequences.

First, in 1876, Nikolaus Otto, Germany, demonstrated the remarkable and enduring 4-stroke internal combustion cycle — now called the Otto Cycle. This first one-cylinder engine produced 500 W (<1 hp), was powered by coal gas, and proved to be sufficiently reliable to power small pumps, lathes, looms, and other machinery; though bulky — and about 1.5 m in height — it sold thousands. Indeed, this Otto Cycle has endured to this day.

Then, in 1886, it happened. Karl Benz, also of Germany, rear-mounted a petroleum burning variation of the Otto Cycle engine on a 3-wheeled platform and connected the engine shaft to the rear axle by bicycle chain; Benz also introduced the differential gear box for turning and a radiator for engine cooling, Fig. 6.3. And it all worked. Then, three years later, fellow countryman Gottlieb Daimler built what is generally considered the first modern automobile: it possessed a V-shaped 2-cylinder Otto Cycle engine, a 4-speed transmission, and produced 1.1 kW (~1.5 hp) at 500 rpm to the crankshaft, Fig. 6.3.

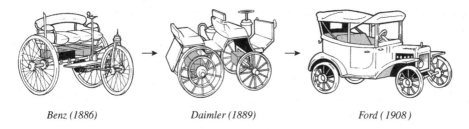

Benz (1886)   *Daimler (1889)*   *Ford (1908)*

Fig. 6.3   Graphical depiction of some early automobile designs.

Before the turn of the century, the Otto/Benz/Daimler automotive developments quickly spread providing also opportunity for related device initiatives. Of particular note are the inventions of John Dunlop, England (pneumatic tire, 1888) and Louis Renault, France (direct transmission, 1899). But it was Henry Ford, USA, who made a particular societal impact. Already in 1896 he had completed his first car, a 2-cylinder 2.2 kW (~3 hp) front-mounted engine but rear-wheel driven by connection with a sprocket chain. In 1908, Ford began component mass production and assembly of his Universal Car using the concept of interchangeable parts manufacture; this rugged Model T, Fig. 6.3, consisted of a wooden body on a steel frame with a 4-cylinder 15 kW (20 hp) engine, was most appropriate for purchasers of modest means but possessing some mechanical aptitude. It was variously improved and when production ceased in 1927, more than 15 million of these low-cost affectionately called *Tin Lizzies* had been produced and other manufacturers were now providing stiff competition. Indeed by the mid-1920s, some 250 firms were manufacturing automobiles worldwide and the Ford dominance was challenged.

This ingenious automobile developmental process may also be compactly depicted by the primal

$$N(t) \begin{Bmatrix} metals, \\ gasoline \end{Bmatrix} \rightarrow E(t) \begin{Bmatrix} mechanical\text{-} \\ thermal\text{-}electro \\ synthesis \end{Bmatrix} \rightarrow D(t)\{automobile\},$$

(6.11)

or, more historically informing, by the connectivity of Fig. 6.4.

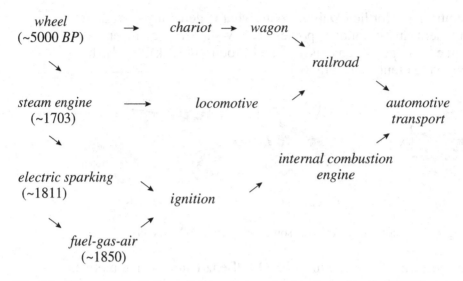

Fig. 6.4   Connectivity diagram showing the progression from the wheel 5000 years ago to the automobile.

Power for Cars

Often forgotten is that in the early automobile era, battery powered and steam powered cars had also been produced with some demonstrating performance features exceeding those of internal combustion powered cars. Range and acceleration as well as differences in manufacturing strategies, seemed to be the main features favoring the internal combustion engine. Now, a century later — as is often the case — past choices are questioned; for engineers, the conjoint issue of *how to make devices* and *how they might be made better* is invariably displaced in time.

6.7   Electrification

The Leyden jar had been invented in 1745 (Leyden, Netherlands) and could store a small quantity of static electricity though the electric charge had to be produced externally by friction and passed into the jar along a conducting rod. Then a device which produced a more convenient though still minimal electrical current was the *pile* built by Alessandro Volta, Italy, in 1800. This first ever battery, consisting of alternate layers of zinc and silver disks separated by brine-soaked paper, quickly spawned a diversity of experiments and prospects for new devices. Already in 1801, this Voltaic pile was used for electroplating and in 1811, Humphrey Davy, England, produced a near-steady arc between two carbon rods to serve as a crude form of lighting; he also observed a faint glow in a current carrying wire thereby foreshadowing the electric light bulb. Then, in 1819, Hans Christian Oersted, Denmark, observed that movement of a current carrying wire would cause a change in alignment of a nearby compass needle, thus discovering the profound at-a-distance effect between an electric current and its magnetic field.

Further critical contributions to the expanded application of electricity were direct descendants of the investigations of the former bookbinder apprentice Michael Faraday, England. In a foundational experiment in 1831, he demonstrated that an electrical conductor moving through the field of a magnet would result in electricity flow in the conductor and, conversely, when an electrical current surged through a conductor a nearby suspended magnet would be deflected. Both principles on which electrical generators and electric motors are based were thus established. Soon after, hand-operated electric current generators in which a horseshoe magnet is rotated near the ends of coils — generally called dynamos — replaced the Voltaic pile as a current source; a commutator was also invented at that time to produce direct current.

Though the Davy-invented arc lamp became popular as a street light, he and others could not eliminate its several shortcomings: harsh light, an operational life of the carbon rods of only a few hours, difficulty of maintaining the appropriate gap between the rods, and the accompanying hissing noise. However, since the Davy observation of a faint glow in a current carrying wire in 1811, it had also been recognized by others that a direct current passing through a sufficiently thin wire might cause incandescence before disintegrating. An inert gas environment or high vacuum was evidently required and such experiments had been undertaken by Joseph Swan of England, in the 1850s, producing incandescence of varying duration. Then, Thomas Edison[†], USA, used a charred cotton filament in a highly evacuated glass bulb and produced a

---

[†]Thomas Alva Edison (1847–1931) was a pioneer in the establishment of research laboratories and is generally considered to be the most prolific inventor with nearly 1100 patents in his name.

continuous glow of light for over 13 hours — the year was 1879. Critical in this development of a practical light bulb was Edison's direct access to various electrical devices and his considerable business acumen.

A strategy for the commercial utilization of electric lighting now became clear: connect incandescent light bulbs in parallel to a DC dynamo powered by a steam engine. Already in 1880, Edison and Swan joined forces to power 3000 light bulbs in London, to the utter amazement of the public. Similar installations were introduced in New York City in 1882. The demand for electrification then became insatiable.

There was however a peculiar limitation: both the voltage at which DC current was generated and the distance of transmission, was limited to about 100 V and about 1 km respectively. Nevertheless, hotels, steamships, factories, office buildings, private mansions, and city-block communities could now have their own steam engine powered dynamos and thereby turning — so to speak — night into day!

It was soon shown that here was another way, prompted by the opportunity to electrify the important railroad and suburban train industry which required longer distances for electricity transmission — but this clearly required AC current. By the 1890s, steam locomotives had experienced extensive improvements but the need for repeated water stops and large quantities of wood or coal, time required to build-up steam pressure, unavoidable noise, and belching sooty smoke, all combined to suggest that perhaps electric motors with AC electricity supplied by overhead wires or a covered extra rail, would be a most desirable improvement in populated areas. This view was promoted by George Westinghouse, USA, and by Wilhelm Siemens, Germany-England. Based on the ideas of the brilliant though irascible Nikola Tesla, Serbia-USA, who had already invented a practical AC generator (1884) and AC motor (1888), Westinghouse promoted AC current because longer distance transmission were feasible. And so began the DC/AC *Battle-of-the-Currents* eventually won by AC, with power from Niagara Falls in 1896 as the first large-scale electric power source.

For the next ~100 years, AC electrification in the USA and Europe — and subsequently Russia and other countries — became an enormous undertaking involving substantial development of water turbines, dam construction, coal fired thermo-electric power plants, high voltage transmission, and step-down voltage transformers. Electricity was first delivered for industrial application, then commercial, then urban-domestic users, and lastly to outlying rural communities, Fig. 6.5. This process of electrification was widely viewed as a dominant expression of 20th century progress largely because of the diversity of application of electrical energy and its general efficiency; it possessed, however, one serious

restriction: it could be stored but only in small quantities, requiring therefore an active and reliable distribution network.

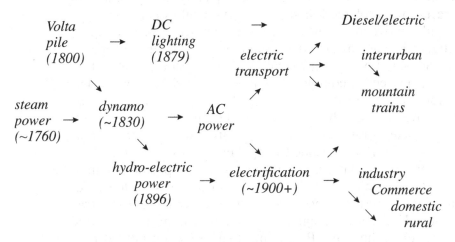

Fig. 6.5   Connectivity diagram illustrating the progression of electric power utilization.

Electrification and Politics

Following the Russian Revolution of 1917 and the resultant leadership of Lenin and subsequently Stalin, electrification became a
highly promoted and very visible political priority in Communist
Russia; this aspiration was often expressed by the slogan

Marxist Socialism + Electrification → Leninism/Communism.

$$\text{(6.12)}$$

There exists little doubt that this large scale program of electrification enabled the phenomenally rapid heavy industry development
of Russia beginning in the 1930s.

But, whatever happened to the prospects of electrical railroads which
provided the initial motivation for long-distance AC electrification? Well,
electrical trains did become operational in urban and mountainous areas
in Europe where population densities and tunnels made steam locomotion unappealing and hazardous, but for long-distance transport, steam
locomotives prevailed until the mid-1900s when diesel-electric locomotives proved to be more satisfactory.

6.8 Electrical Communication

The invention of the Voltaic battery and subsequently the dynamo provided a treasure trove of modest-scale experimental opportunities for inquisitive and creative minds. Many of these innovations proved highly consequential to the ensuing electrical communication industries of telegraphy, telephony, radio, and television. But first a historical context.

The contrast in speed of communication prior to the invention and use of the telegraph in the 1830s is astounding. Recall that until then messages could move no faster than a person could run, ride a horse, travel by stagecoach, take a train, or sail a ship. But there was one remarkable exception: in the 1790s Claude Chappe, France, experimented with adjustable colored wooden boards on a tower, with their positioning indicating a coded message readable by an observer on a distant tower for similar transmittal to the next tower, and so on; note its analogue with past North American Native smoke signals. Then, a paradigm shift occurred in communication technology. But, rather than based on optical phenomena, it was the ingenious use of electromagnetic phenomena which provided a most profound change in the means of communication. This change began in the early 1800s and continues even now, 200 years later.

6.8.1 *Telegraphy*

Soon after Volta's pile invention, it was recognized that disruption of an electric current in a conducting wire would propagate unchanged for some distance. This property of electrical conductivity suggested that coded information might be carried by wire, requiring however that effective devices be developed which might quantitatively measure these current disruptions. A range of experiments were undertaken which, at one extreme involved electrodes immersed in an electrolyte with bubbles forming in proportion to the current, to the more practical tests involving the response of a freely suspended magnetic needle near the conducting wire. Continuing investigation along the latter line of inquiry led to the galvanometer and the telegraph.

An early telegraph had been patented in England in 1837 and consisted of a 6-wire battery-driven closed circuit with each letter of the alphabet coded to a current flow in a prespecified number of the six conductors. This system, though cumbersome, worked well over distances of ~100 km and the burgeoning railway industry, having a convenient right-of-way for stringing electrical cable and needing to keep their station masters informed of railway movement[†], showed an immediate interest.

Also in 1837, Samuel Morse, USA, proposed a simpler dot-dash code for the alphabet which could be incorporated into a single wire circuit. It could also be driven by a Voltaic battery and the current would be disrupted by a hand operated tappable key; a corresponding tapping solenoid and diaphragm at the receiver would audibly reproduce the signal. This simple Morse code system found immediate acceptance by the news media and by some commercial establishments so that telegraph lines soon criss-crossed Europe and the USA along the rail lines and into nearby settlements. An undersea cable of the English Channel was placed in 1851 and intercontinental telegraphy followed with the first transatlantic cable in 1866. Significant improvements were continuously introduced, including the teletypewriter in 1855 and multiplexing in 1875. Telegraphy remained a dominant means of long-distance communication for about a century.

6.8.2 *Telephony*

With telegraphy a success, many creative minds envisaged the possibility of similarly transmitting human speech along a conducting wire. Of various experiments undertaken in Europe and North America, it was the device patented in 1876 by the speech instructor Alexander Graham Bell, Scotland-Canada-USA, which succeeded. Critical to this invention was the vibrational spectrum of a parchment covered granular diaphragm when exposed to speech sound waves. These vibrations would affect the magnetic field of an adjacent electromagnet and induce a corresponding current flow pattern in the battery-driven conducting wire. An inverse process would occur at the receiver's terminus thus reproducing the originating audible sound.

In 1878 the first commercial exchange with about 20 subscribers became operational in Boston and in 1884 Boston and New York were linked. The number of telephone installations had grown rapidly and within 100 years it had increased to about 500 million worldwide. Interestingly, the Bell invention has also proven to be remarkably adaptable for myriad continuing improvements.

[†]This also introduced the idea of standardization of time over extended regions, but was soon subverted when different railroad companies introduced their own unique reference t_0 time coordinate. Universal Standard Time, introduced in 1883, put this annoyance to rest.

Invention: Simultaneity and Recognition

When Alexander Graham Bell filed his patent for the telephone, a man named Elisha Gray also arrived at the Patent Office with a similar invention, but 2 hours later. Because of this delay, Gray lost his place in history, for the telephone is widely recognized as the most lucrative single patent ever obtained. Interestingly, the importance of the telephone had not been recognized at the time for soon after being granted the patent, Bell — it is said — offered it to the communication giant Western Union for a modest sum. Western Union declined, reasoning that in view of its successful telegraph, a telephone — for its distance limitations of the day — was of little relevance. A year later Western Union changed its mind, but so had Bell.

6.8.3 *Radio*

Even before the demonstration of radiowaves had been established, James Maxwell, Scotland, had in 1865 formulated a mathematical characterization of the known electrical and magnetic dynamic phenomena which pointed to the existence of an invisible form of radiation propagating through space at the speed of light. Maxwell's idea now suggested a profound extension of the telegraph and telephone: if an electrical phenomenon transmits information in a conducting wire, might not electromagnetic waves also transmit information in the earth's atmosphere? And thus, one may assert, the radio was the first modern device conceived in theory before a working device was demonstrated in practice.

In 1887, Heinrich Hertz, Germany, provided the first laboratory demonstration of such atmospheric electromagnetic propagation and reception: a resonating circuit was designed to generate a spark between two metallic spheres and the consequent electrical disturbance was transported in air and detected by a nearby conducting loop containing a small gap. A profound long distance application of this concept of spark-gap induced electromagnetic wave propagation was demonstrated in 1901 by Guglielmo Marconi, Italy-England, when a coded dot-dot-dot electrical signal was triggered in England and received in St. John's, Newfoundland. The era of wireless telegraphy transmission, amplification, receiving, and telecommunication, then began with numerous competing inventions and innovations. And then, in 1906, Reginald Fessenden, Canada-USA, invented continuous-wave signal transmission and reception, thereby introducing both the radio and the radio-telephone. This invention led to commercial radio and, decades later, provided the basis for television and the wireless telephone.

The Threat of Open Communication

The appearance of a means of transmitting information through the earth's atmosphere generated considerable fear among some governments, and for good reason — no paper to confiscate, no transmitting wire to cut, no messenger to hold or restrain. They sensed that this constituted a powerful and penetrating means for potential subversive activity. As a consequence, some governments became controlling owners of such installations while others introduced tight regulations on this new form of communication.

6.8.4 *Television*

Radio involves the transmission of one-dimensional audio information superimposed on an electromagnetic carrier wave. By analogy and extension, the concept of transmitting spatial shadings of grayness associated with two-dimensional images became of wide interest. Devices for the scanning and reproduction of planar grayness information now became a focus for inventions. The several and arguably most important perspectives and device developments can be summarized as follows:

1884: Paul Nipkow, Germany, introduced a spiral disc which reduced an image to individual pixels of varying levels of grayness.

1923: Vladimir Zworykin, Russia-USA, developed an optical scanning device in which an image is projected onto a screen for continuous and sequential scanning by a narrow electron beam; electron reflection being proportional to grayness thus provided an electronic spatial intensity analog of the source image.

1926: John Baird, Scotland, used Nipkow discs to produce television images.

1927: Philo Farnsworth, USA, constructed an electronic image scanning and transmission device.

1939: Beginning of commercial television in the USA.

Note here the ~50-year history of development from inception to commercialization, followed by decades of continuing device improvement and expanded uses.

And so a new primal progression appropriate to electrical communication emerged:

$$N(t) \begin{Bmatrix} various \\ materials \end{Bmatrix} \rightarrow E(t) \begin{Bmatrix} EM\ transmission \\ and\ detection \end{Bmatrix} \rightarrow D(t) \begin{Bmatrix} telegraph, \\ telephone, \\ radio, \\ television \end{Bmatrix},$$

$$(6.13)$$

and all based on the theoretical musings of James Maxwell — some theories can indeed be most practical.

6.9   Air Transport

Moving through the air has long been an exceptional subject of immense fascination. A Chinese legend dating to about 1500 BCE tells of *Ki-King* traveling in a flying chariot with the wheels serving as rotating wings. Icarus of Greek mythology escaped from the island of Crete on wings made by his father Daedalus, but he came too close to the sun which melted the wax on his wings and he crashed to earth. The first Japanese Emperor is believed to have descended by a flying ship, popular folklore has it that Eastern Sultans once soared on flying carpets, and Native Americans have many stories of escaping imminent danger in the claws of thunderbirds.

More practical visionaries like Leonardo da Vinci conceived of a helicopter and a parachute and Joseph and Jacques Montgolfier of France actually did rise in hot air cloth-and-paper balloons in 1783; indeed, helicopters and parachutes were eventually built and recreational hot air ballooning did become of sporting interest.

A particularly noteworthy contribution to heavier-than-air devices for flying can be attributed to a multitalented English aristocrat named George Cayley. In the early 1800s, he assembled and further determined much useful information on flying, such as wing area needed to support a given weight, the role of air resistance, stability and control requirements, and eventually specified the need for a power source. Not having such an engine available, he nevertheless did build a marginally effective glider consisting of a 3-wheel platform suspended from two large wings.

Cayley also came to the remarkable conclusion that a fish swimming in water was a better simulation model for an aircraft than a bird flying in air.

The Reluctant Aviator

A popular anecdote has it that Cayley persuaded his elderly and very formal coachman to pilot a crude glider. Not having the fortitude or adventurous spirit of future aviators, the coachman staunchly resisted but finally did consent. After a distance of ~100 m of haphazard gliding, he stumbled out in a daze and stuttered to say that he was hired to drive a coach and not to fly an aeroplane. He also quit his coachman job in protest.

Beginning in the mid-1800s, a number of flying enthusiasts experimented with various designs, including even steam engines as power sources. However, the demonstration of powered flying had to await the hang-glider experiments of Lillienthal and the careful design and systematic testing of the Wright Brothers.

During the period 1891–1896, Otto Lillienthal, Germany, built various types of shoulder-and-arms harnessed hang glider arrangements which required running downhill to become airborne and by body movement achieve some flight control. He performed about 2000 flights ranging up to 400 m in length and in 1896 began work on a powered glider, proposing to use one of the recently developed internal combustion engines. However, he died as a result of injuries from a glider mishap before this new project could be completed. Evidently, Lillienthal was the first human to spend significant time airborne and may well be considered the world's first hang glider.

The enormous achievement of first controlled powered flight was accomplished by brothers Wilbur and Orville Wright, methodical and highly motivated bicycle mechanics from Dayton, OH. They knew about the Lillienthal experience and in 1900 began assembling and testing large cloth covered wood and wire frame kites and gliders at a windy and sandy barrier island near Kitty Hawk, NC. As a consequence of numerous aerodynamic tests, they developed the warped-wing airfoil concept, introduced a forward canard for vertical control, rear rudders for turning, and began construction of their powered aircraft in 1902. This was a 300 kg and 13 m wingspan biplane, equipped with a 12 kW (16 hp),

4-cylinder, gasoline powered internal combustion engine of their own design and manufacture, with two rear mounted contrarotating propellers connected to the engine crankshaft by bicycle chains. Wooden skids were mounted for take-off from a 20 m wooden monorail, requiring therefore a good head wind. Full-scale experiments were scheduled for the autumn of 1903.

On the first attempt in December 1903, their aircraft accelerated along the launching rail, rose to about 3 m, then stalled, and mildly crashed. The damage however was slight and quickly repaired. Three days later, 17 December 1903, head-wind wind conditions were more favorable and the plane rose to a height of 4 m and 12 s later landed 40 m from the point of take-off. Three more flights were undertaken that day with the last staying aloft for almost one minute and covering a distance of nearly 300 m. The first ever manned, powered, and controlled flight was accomplished, witnessed by four other adults and two children.

The Wright Brothers' pioneering demonstrations contributed to the onset of a European and American aircraft testing, construction, and flying craze so that by 1906 several improved aircraft were flying — including improved versions of the Wright Brother's design. At the first ever air show in Paris in 1910, 40 different aircraft from 10 different manufacturers were on display and in the same year governments started acquiring aircraft for military purposes. And thus was born the phenomenal growth industry suggested by the primal

$$N(t)\{metals, petroleum\} \rightarrow E(t)\{aeronautics\} \rightarrow D(t)\{aircraft\}. \tag{6.14}$$

We need to extract from this heavier-than-air flight history the case of lighter-than-air flying, that is extensions of the Montgolfier 1783 hot-air ballooning experiments, to large hydrogen or helium-filled airships with a gondola suspended below for crew and passengers. In the early 1900s, several countries sponsored construction of such elongated cigar-shaped airships, 100 to 250 m in length. Vertical motion was attained by discharging ballast or excess lifting gas and horizontal motion was attained by pusher propellers geared by crankshaft to internal combustion engines located in suspended pods. Publicity about round-the-world airship travel generated much attention and in 1919 regular trans-Atlantic travel was initiated. Though stormy weather continued to be an operational hazard, the commercial airship future came to a dramatic end in 1937 when the 72-passenger hydrogen-filled Hindenburg airship burst into flames while docking at Lakehurst, NJ.

Some noteworthy reference events associated with the consequent development of air transport are the following:

1919: crossing of the Atlantic (5-man USA crew, refueling in Newfoundland and Azores)
1927: solo non-stop crossing of the Atlantic (Charles Lindbergh, 33 hours)
1928: rocket propelled aircraft testing
1933: begin of commercial air transport
1939: jet aircraft flight demonstration
1942: operational helicopter
1953: begin of commercial jet transport

And so the imaginative writers of sky-transport legends had their dreams become reality with engineers as the ingenious facilitators.

6.10 Heavy Industry

The progression of various processes for the production of quality steel

$$\begin{array}{ccccc} Bessemer & & Open\ Hearth & & Electric\ Arc \\ (\sim 1856) & \rightarrow & (\sim 1880) & \rightarrow & (\sim 1920) \end{array} \rightarrow \cdots \qquad (6.15)$$

became exceedingly successful and this for three reasons:

(a) Smelters could be readily expanded with few limitations.
(b) From ~ 1860 to ~ 1890, cost of steel production decreased by a factor 10.
(c) Demand for steel increased continually to the mid-1900s.

Applications of quality steel were associated primarily with specific industries requiring increasingly expanding device production:

(a) *Railways*
   By ~ 1900, railways were among the largest industrial employers in Europe and North America with a steadily rising need for steel.
(b) *Military*
   World War I initiated a considerable arms race involving steel for heavier tanks, larger warships, and more lethal small arms.
(c) *Toolmaking and Machinery*
   Increasing mass production by the turn of the century demanded durable tools and precision machinery, requiring therefore increasing production of specialty steels.

(d) *Structures and Transportation*

With the Eiffel tower built of steel for the Paris Exhibition of 1889, subsequent tall structures — skyscrapers, large domes, and large bridges — also required special structural steel and steel cables, and eventually steel reinforced concrete. The expanding automotive, farm implements, and shipping industry also benefitted from specialty steel and various alloys.

Steel became widely produced metal during the latter half of the Expansive Engineering period. However, its large smelting, processing, and refining installations demanded considerable power — produced primarily from the burning of coal — and inevitably leading to steelmaking becoming associated with belching smokestacks, large piles of exposed slag, and extensive environmental pollution.

The Works of Carnegie

The prime contributor to the USA emergence as the dominant global steel producer was a poor Scottish immigrant named Andrew Carnegie (1835–1919). Starting in the 1860s with a small iron works in Pittsburgh and a singular vision of the emerging importance of quality steel, he soon became its leading producer and one of the wealthiest individuals, eventually endowing the Carnegie Institute of Technology in Pittsburgh, the Carnegie Institute of Washington, the Carnegie Concert Hall in New York City, and — as a most unusual but very consequential act of philanthropy — nearly 3000 community libraries in the USA and Canada.

Considerations of environmental pollution makes it essential to distinguish between natural processes which produce substances hazardous to humans — volcanic ash, decaying vegetation, and decomposing animal matter — from anthropogenic (i.e. human caused) releases of toxic and contaminating substances. It has been estimated that by the mid-1900s, anthropogenic production of toxic and polluting substances exceeded those produced naturally. The heavy industry, electricity producing coal-fired generating plants, and the internal combustion powered transportation industry, became the dominant producers of hazardous anthropogenic substances emitted into the environment.

6.11 Synthetic Materials

We had previously noted that the late 1800s were characterized by four material transformation processes of particular significance to engineering: Bessemer steelmaking, synthetic textile dyes, vulcanized rubber, and stabilization of nitroglycerine. With the advantage of hindsight, one may well assert that of these four it was the underlying polymerization feature of vulcanized rubber which has been particularly consequential for the 20th century; indeed, a tentative step towards polymeric materials had already occurred with the invention of celluloid in 1862, though its uses had proven to be limited.

By the 20th century, science had identified many basic classes of materials. While atoms were the basic building blocks combining in myriad ways to form molecules, it then also became important to make the following distinctions:

(a) *Monomers*
    Small molecular assemblages of simple substances (e.g. H_2O, NaCl, ...) called monomers.
(b) *Polymers*
    Large hydro-carbon based assemblages composed of repeating strongly cross-connected molecular units called polymers and characterized by $\sim 10^6$ atomic components and even more.

Nature seems to have a special interest in polymers. For example, all plants and animals possess numerous kinds of polymers such as glucose, cellulose, proteins, and amino acids; indeed, natural rubber and some natural glues extracted from some trees are polymers and possess some most exceptional properties: they can be elastic, rigid, or pliable under some conditions of temperature and the attainment of particularly desirable properties might be established by a judicious choice of mixing and processing. In the early 1900s, many entrepreneurs and some corporations developed an interest in polymers, especially in the manufacture of compounds which could yield particular devices such as longer-lasting tires, fire resistant clothing, and durable insulators.

The first such 20th century polymer of importance was Bakelite, discovered by the chemist Henrik Baekeland, Belgium-USA, in 1909. This polymer is a thermosetting resin composed of coal tar condensates phenol and formaldehyde. It was suitable as a nonflammable electrical insulating material for the continuing growth of electrification.

Some large corporations soon established research and developmental centers specifically for the development and production of useful polymers. In 1926, B.F. Goodrich, USA, produced Polyvinylchloride (PVC), an easily colored substance useful for flexible coatings, molded products, and various types of tubings. Then, the German firm I.G. Farben produced a foam called Polystyrene in 1930 which became widely used for insulation, packaging, and flotation purposes. Also, a tough light substitute for glass, called Plexiglas, became commercially available in 1934.

Probably the most prolific developer of commercially useful polymers was the USA chemical giant duPont, formally known as E.I. duPont de Nemours and Company. Its chief polymeric researcher Wallace Carothers promoted a fundamental approach to the study of polymers and this emphasis soon resulted in a series of successful polymeric products: in 1931 Neoprene, a good substitute for vulcanized rubber; in 1937 Lucite, a competitor to Plexiglas; and in quick succession Teflon, Polyethylene, Rayon, Acrylic, and other products. A particularly noteworthy product was Nylon, a silk substitute with exceptional properties of fibrous strength. Nylon went into production in 1938 and for the next several years its output was largely directed to war-related products such as parachutes, bearings, tire cord, flak vests, rope, and a small production allocation for fishing lines, toothbrush bristles, and — the ultimate fashion product of the day — sheer stockings! Thus, the primal progression

$$N(t) \left\{ \begin{array}{c} natural \\ materials \end{array} \right\} \rightarrow E(t) \left\{ \begin{array}{c} chemical \\ synthesis \end{array} \right\} \rightarrow D(t) \left\{ \begin{array}{c} polymeric \\ products \end{array} \right\} \quad (6.16)$$

encapsules this important process of synthetics which has enormously contributed to superior performance of innumerable devices of wide consumer interest.

Common/Uncommon Materials

Divers exploring the recently discovered Titanic which sank in 1912, were surprised by the absence of a most common present-day material: the complete absence of plastics. Metal, wood, leather, fabric, quantities of decorative stone, and — as a sign of lavish opulence — quality porcelain and ornate silverware were found, but no plastics!

6.12 Large Structures

The Expansive Engineering era experienced the continuing extension of a range of structures concurrently with the development of land, air, and sea transport and an expanded network for water and energy distribution.

Of initial importance in this extension were canals. While canal construction for irrigation purposes dates to Ancient times and canals for inland transport dates to Renascent times, canals as a means of oceanic distance-of-travel reduction became of international interest beginning in the mid 1800s. Most noteworthy of such civil construction projects was the 160 km Suez Canal, constructed over the 10-year period 1859–1869, and the 80 km Panama Canal, intermittently built during 1894–1914.

Bridge construction began to flourish in tandem with railroad development in the early 1800s, with much variation in detailed design but mostly consisting of wood truss, iron cantilever, and iron arches. These early bridges possessed initial spans typically up to ~200 m and increased to ~500 m by the early 1900s. An exceptional pioneering cantilever multipier bridge crossed the 1500 m Firth of Forth in Scotland in the 1880; this remarkably imposing engineering structure soon became known as the Sight of Europe and even attracted royalty and other notables as amazed tourists. But the most publically impressive bridges were the family of suspension designs: the pioneering Brooklyn Bridge with a span of 500 m, completed in 1883, while the graceful Golden Gate Bridge has a span of 1300 m and opened to the public in 1937.

Less visible but very demanding construction projects of the late 1800s were the various subways and tunnels. Already in 1863, London began to operate the first stage of its Underground *Tube*, New York's similar *Subway* became operational in 1868, and Paris began building its *Metro* in 1898. Heavy railroad traffic tunnel usage began with the 14 km Mont Cenis Tunnel in the Alps in 1870 and underwater vehicular travel began in 1920 with the 3 km Holland Tunnel under the Hudson River in New York City.

A remarkable set of multipurpose construction activities of the Expansive Engineering era involved the functional-triad of electricity generation, flood control, and irrigation water supply, all provided by one dam situated across a suitable location of a sufficiently flowing river. Initially, low earth-fill and then concrete-arch dams were constructed, eventually leading to some exceptionally large dams. The highest concrete dam of this era was the 250 m high concrete-arch Hoover Dam across the Colorado River, completed in 1936, and the largest concrete-volume dam of that time is the 1300 m long Grand Coulee Dam across the Columbia River, completed in 1940.

Other noteworthy structures of the Expansive Engineering era included the 300 m Eiffel Tower completed in 1884 and the 420 m Empire State Building which opened in 1931. In a class of their own are the limited access Auto Routes, Autobahns, and Turnpikes beginning in the 1930s; simultaneously, a rapidly growing petroleum distribution network appeared involving the integration of ocean tankers, pipelines, trucks, and large storage tanks.

The Birth of Freeways

In 1925, the 25 km Bronx River Parkway introduced the idea of a divided highway and in 1928 a cloverleaf exchange was built in New Jersey. Then in 1940, the 250 km Pennsylvania Turnpike opened to public traffic with no posted speed limit. The USA Interstate System of Freeways was begun in 1953 and presently extends to ~80,000 km in length; in every case, curves and grades were designed according to specific safety and economic criteria.

6.13   Engineering Organizations

Until the early Renaissance, the actual workings of engineers was largely pursued as an individualistic activity. Pride in workmanship, ambition to see one's devices take shape, and the recognition of patrons and the public were the private aspirations of the engineer. An initial sense of group identity became established in the 1400s when engineering manuals and handbooks on successful design and construction experience became available; this process of technical information dissemination was subsequently hastened by the Gutenberg printing press. The ensuing Industrial Revolution tended to identify the makers of ingenious devices as highly specialized crafters as much as engineers.

In the late 1600s and 1700s, it was the French advanced-level programs in military engineering, transportation engineering, hydraulic engineering — and the pace-setting Polytechnics — which tended to identify a cadre of engineers characterized by a shared technical knowledge base and common objectives.

But as is so often the case, the transition to a more professional sense of engineering as-a-community identity was triggered by an individual: John Smeaton of England.

John Smeaton (1724–1792) chose engineering as a career by way of an apprenticeship in instrument making. By further self-study and careful investigations he became a remarkably accomplished engineer of his day. He improved on the steam engine, designed a superb lighthouse, became an effective expert witness in courts of law, and was among the first to call himself Civil Engineer — as a distinction to that of a Military Engineer. Beginning in 1771, he organized dinner-discussion meetings of like-minded engineers and this group soon called itself the Society of Civil Engineers. In 1828, they became incorporated as the Institution of Civil Engineers.

Engineering organizational initiatives in the USA were stimulated in the early 1800s by expanded road and bridge construction, the large Erie Canal project, and the first university level engineering programs at Norwich University, Norwich, CT, in 1821 and at Rensselaer Polytechnic Institute, Troy, NY, in 1824. The first engineering association was the American Society of Civil Engineers (ASCE) founded in Chicago in 1852 and other engineering associations soon followed: American Institute of Mining, Metallurgical, and Petroleum Engineering (AIME) in 1871, American Society of Mechanical Engineers (ASME) in 1880, Institute of Electrical and Electronic Engineers (IEEE) in 1884, American Institute of Chemical Engineers (AIChE) in 1908, and subsequently many others.

The principal objective of these engineering associations was and continues to be the dissemination of technical information largely based on publications and conferences. It is important to note that these and similar engineering organizations are not licencing agencies; such functions are the responsibility of regional boards.

6.14 Expansive Engineering: Environmental Impact of Devices

In Chapter 5, specific reference was made to the environmental impact of the expanded smelting and metal refining processes associated with the stimulating effects of the Industrial Revolution. This environmental impact became considerably intensified during the Expansive Engineering era, and this for two reasons:

(a) *Power/Energy Production*
   In addition to the continuing and expanded burning of coal, much of the increasing demand for energy also became supplied by the

combustion of petroleum products. Whether burning wood, coal, natural gas, or petroleum, a dominant hydro-carbon combustion process[†] is suggested by

$$(CH_4 + various\ contaminants)_{fuel} + (2O_2 + various\ components)_{air}$$
$$\rightarrow 2H_2O + CO_2 + \begin{pmatrix} contaminant \\ products\ NO_x, SO_2, \ldots \end{pmatrix} + \begin{pmatrix} thermal \\ energy\ gain \end{pmatrix}.$$

$$(6.17)$$

Thus, gases CO_2, NO_x, SO_2, and other particulate contaminants are released into the environment. The consequences suggested above — that is reaction products and thermal energy — are specifically associated with the $N(t) \rightarrow E(t)$ part of the engineering progression, thereby expanding the contamination flow first noted in Eq. (5.17).

(b) *Device Manufacture and Uses*
Further, the increasing production of devices — that is the manufacturing processes associated with the $E(t) \rightarrow D(t)$ part of the engineering core, as well as the increasing use of transportation devices by society implied in the $D(t) \rightarrow S(t)$ term — also generates particulate and waste energy as a similar contributor to environmental degradation. Hence, the existing engineering progression needs to be augmented as

Environmental Contamination

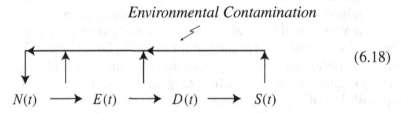

$$(6.18)$$

Finally, this Expansive Engineering era also led to an increasing accumulation of devices commonly judged to be spent, obsolete, or sufficiently worn — and hence of little further use. Typical examples are worn transport tires, damaged and obsolete military equipment, automobile wrecks, old household appliances, obsolete apparel, outdated factory machines, dated newspapers and magazines, and many others. These wastes or disposed devices may be accounted for by the identification of

---

[†]Depending upon the source, some 70% to 98% of the molecular composition of gaseous and liquid hydrocarbon fuels is methane, CH_4.

an accumulation of sequestered waste or remains inventory, to be labelled the repository $R(t)$, and incorporated by an addition to the core of the heterogeneous connectivity progression (6.18):

$$N(t) \rightarrow E(t) \rightarrow D(t) \rightarrow S(t) \rightarrow R(t). \qquad (6.19)$$

In practice, $R(t)$ constitutes a widely distributed set of locations including landfill sites for household garbage, junk piles of spent consumer goods, slag piles at mining locations, toxic pools near processing plants, hazardous concentrates at manufacturing facilities, heaps of worn tires at desolate locations, scrap yards at factories, sunken ship hulks, regional material storage facilities, graveyards of obsolete military aircraft at isolated landing strips, etc.

As a further comment, note that to the end of the preceding Renascent Engineering era, the repository inventory could largely be neglected because most of the devices were often made of decomposable wood and natural fiber which eventually entered the natural flow of materials; and those which did not fully decompose, such as iron devices or obsolete metal farm implements as well as failed or spent machine parts, represented a relatively small and diluted form of contamination. This, however, is not to detract from the recognition that both environmental contamination and non-decomposable waste accumulation had their substantial beginning in the Industrial Revolution, beginning $\sim$1760.

This additional detail now provides for the Expansive Engineering network by the integration of the above as additions to Eq. (5.18)

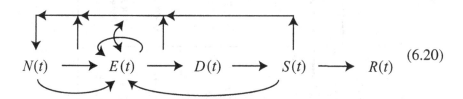

$$(6.20)$$

Note that the straight arrows primarily represent material and energy flows and the curved arrows are to suggest information and ingenuity flows.

Table 6.1 brings our sequence of engineering up to the end of the Expansive Engineering era.

Table 6.1   Addition of the Expansive Engineering period ($\sim$1800 → 1940) to the evolution of engineering. Note again the expanding characterization when viewed as an evolving non-symmetric matrix.

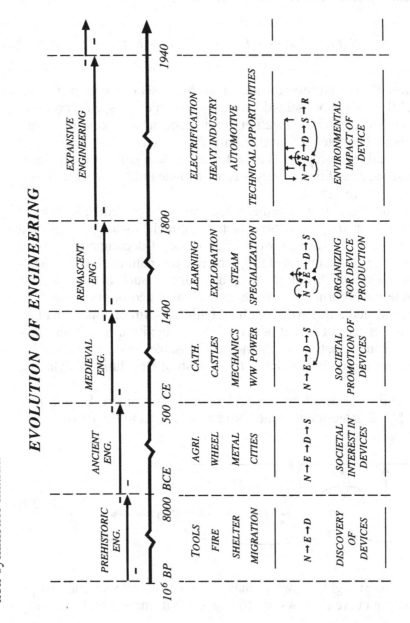

6.15 To Think About

- The Expansive Engineering era ($\sim$1800 $\rightarrow$ $\sim$1940) is sometimes viewed as a highly opportunistic engineering period having given rise to a disproportionate heavy industrial production and a wasteful material-energy culture. Discuss.
- Common expression such as *explosion of technological knowledge* suggests unboundedness. Is this idea reasonable? Do some boundaries exist? Explain.
- It appears that a significant part of device manufacturing skill and technical knowledge seems to flow from East to West: Middle East $\rightarrow$ Europe $\rightarrow$ North America $\rightarrow$ Asia. Discuss this pattern and research some exceptions and possibly reversions.

CHAPTER 7

Modern Engineering

($\sim$1940 $\rightarrow$ $\sim$1990)

Expanding Reach of Devices

7.1 New Engineering Panoramas

Significant changes in engineering theory and practice characterize the emergence and ultimate features of Modern Engineering,
$\sim$1940 $\rightarrow$$\sim$1990. While some of these changes represented traditional
forms of continuing extension and specialization of devices, it also
became evident that engineering of this era evolved with an increasing
scientific focus, projecting additionally an expanded interest in meeting societal expectations. We consider selected aspects of engineering developments of this Modern period proceeding by decadal time
increments.

7.1.1 *The 1940s: Nuclear Energy*

In December 1938, a profound discovery in nuclear physics occurred
whose subsequent engineering manifestation was to lead to a significant global political impact. In this discovery it was shown that the
Uranium-235 nucleus (92 protons, 143 neutrons) could be split with
an attendant per-reaction energy release about 10^7 greater than that
achievable by conventional hydrocarbon-based combustion. Additionally, the progression of this enormous energy release could decline,
propagate steadily, or grow explosively, by neutron-driven auto-catalytic
process in a medium containing a sufficient volume and density of this

fissile U-235 isotope; this self-chaining reaction cycle may be represented by

$$^{235}U \nearrow \quad \begin{array}{c} \nearrow \: ^{235}U \searrow \\ \\ \\ \searrow \: n \: \nearrow \end{array} \quad \searrow \: \rightarrow \: \sim 1.2 \: \textit{neutrons} \qquad (7.1)$$

highly energetic heat-producing
reaction product

Thus, in addition to the molecular fossil-fuel combustion process, (6.17), scientists and engineers had identified another exceedingly exoergic nuclear reaction.

Two military consequences promptly ensued. One was the development of nuclear explosives with a first full-scale test in July 1945 in New Mexico, USA, followed by its use in WWII a month later. The other was the important recognition that since this nuclear chain reaction did not require oxygen as a reactant, it could constitute an ideal power source for submarines. By the early 1950s, the first such nuclear submarine entered service, astonishing even its promoters by its superior submerged speed and range. Other nuclear submarines and also nuclear powered aircraft carriers soon followed.

The technical success of the submarine nuclear power plant also suggested a reasonable extension: a scaled-up and modified naval nuclear reactor to be used as a power plant to generate electricity for the civilian electrical energy market. The ensuing fast track program of military-to-civilian development in several countries led to the first such nuclear-electric power plants becoming operational in the mid 1950s, in the USSR and USA, followed by some 400 large ($>10^3$ MW) nuclear-electric power plants world-wide during the next 25 years.

Both naval vessels and civilian electricity generators use the same type of nuclear reaction and similar device configuration but with the released energy used for different purposes:

$$\begin{pmatrix} \textit{energetic} \\ \textit{nuclear} \\ \textit{products} \end{pmatrix} \rightarrow \begin{pmatrix} \textit{heating of} \\ \textit{water to} \\ \textit{form steam} \end{pmatrix} \rightarrow \begin{pmatrix} \textit{steam} \\ \textit{turbine} \end{pmatrix} \begin{array}{c} \nearrow \textit{power for military propulsion} \\ \\ \searrow \textit{power for civilian electricity} \end{array}$$

$$(7.2)$$

This kind of military-civilian progression is often referred to as *dual-use* technology.

Primary Inanimate Energy Sources

Prehistoric engineers discovered how to start and control wood fire; the primal basis for these processes is the thermally-driven rearrangement of some hydrocarbon molecules found in all vegetation, and their autonomous exoergic reaction chaining under commonly encountered conditions. Successive engineers expanded on this process to include solid (coal), liquid (petroleum), and gaseous (natural gas) forms of hydrocarbon burning. Modern engineering introduced a paradigm shift in energy production by the process of nuclear rearrangement in the naturally occurring heavy isotope Uranium-235 (i.e. fission energy) and postcontemporary engineers are expected to rearrange the nuclei of light elements (i.e. fusion energy) as a further energy supply alternative. Each of these three primary conversion processes offers distinct variations on availability of fuel supply, ignition conditions, control requirements, and environmental side effects.

7.1.2 *The 1950s: Computation and Computer*

The electronic computer emerged in the 1950s as a most visible and increasingly important device, providing initially a centralized computation and simulation support capability for research and industry. Its use then expanded until no part of contemporary existence escaped the impact of this pervasive and multi-faceted ingenious device. Indeed, if the steam engine of the 1700s triggered a *Hard Revolution*, the computer of the 1950s triggered a *Soft Revolution*.

At their most foundational level, present-day computers together with various electronic calculators and microprocessors, perform a type of coded clay-tablet accounting and communication function first used by the prehistoric Sumerians and which also appeared in sliding-beads-on-wires arrangement of ancient Asia, finally culminating in the Chinese abacus, ~300 CE. Following these ancient inventions, the next significant device development — occurring 12 centuries later — was the discovery by John Napier, Scotland, that multiplication/division was equivalent to adding/subtracting corresponding exponents and this was quickly followed by the invention of the slide rule by William Oughtred, England, enabling these calculations to be simple and fast. It was however the genius of Blaise Pascal, France, in the mid 1600s, which yielded the first mechanical adding and subtracting machine. A more ambitious mechanical calculator — involving also the innovation of punched wooden blocks for data input — was devised by Charles Babbage, England, in 1822. The era of electronic computers had its embryonic beginnings with the idea of vacuum tube binary electronic coding by John Atanasoff, USA, in 1937, followed by the Electronic Numerical Integrator and Calculator (ENIAC) in 1945 at the University of Pennsylvania, USA. Contemporary computers also trace their development to the critical ideas of Alan Turing, England, and John von Neumann, Hungary-USA; the latter introduced the concept of a stored program of instructions which was first incorporated in 1950 in the Universal Automatic Computer (UNIVAC). Thus the heterogeneous progression

$$
\begin{array}{ccccc}
\textit{sliding-beads} & \textit{sliding-bead} & \textit{slide} & \textit{mechanical} & \\
\textit{-on-wires} \rightarrow & \textit{abacus} \rightarrow & \textit{rule} \rightarrow & \textit{computer} \rightarrow & \cdots \\
(\sim 5000\ BP) & (\sim 300\ CE) & (1620) & (1649,\ 1822) &
\end{array}
$$

$$(7.3)$$

compactly encapsules about 7000 years of data management and computation by various mechanical devices.

A nearly continuous stream of various types of electronic computers then emerged and proved useful for a range of computation, data management, and communication applications: scientific, industrial, business, and eventually for personal uses. Computer research and development has been intensively pursued, largely with three broad areas of emphasis:

(a) *Hardware*
    This developmental emphasis involved the device progression

$$
\begin{array}{cccc}
\textit{electronic} & \textit{solid-state} & \textit{integrated} & \textit{large integrated} \\
\textit{tube} \rightarrow & \textit{transistor} \rightarrow & \textit{circuit} \rightarrow & \textit{circuit systems} \quad (7.4a) \\
(1906) & (1947) & (1958) & (1970)
\end{array}
$$

Remarkable success has been possible, leading to increased memory capacity, size reduction, computational speed, and lower cost.

(b) *Software*
    Software has evolved with its own unique progression

$$
\begin{array}{cccc}
\textit{binary} & \textit{stored} & \textit{formula} & \textit{multi-tasking} \\
\textit{coding} \rightarrow & \textit{programs} \rightarrow & \textit{translation} \rightarrow & \textit{operating systems} \quad (7.4b) \\
(1940s) & (1950s) & (1960s) & (1970s)
\end{array}
$$

and continues.

(c) *Firmware*

This involves micro devices in which a specific set of software programming capability is embodied in hardware design. These microprocessors have in recent years found an important niche in assembly line manufacturing, automobiles, and household appliances.

By any measure of assessment, the computer has become an important device for engineering purposes and has influenced human practices in most profound ways.

Mainframe versus PC Strategies

First generation electronic computers were large centralized facilities requiring therefore that customers in need of computation had to travel some distance. A common urban myth has it that an *authoritative* opinion of the day was that in view of the expense of such large centralized high-capacity computational installations, the market could not accommodate more than perhaps *a dozen* mainframe installations worldwide; factual or not — and probably not — the development of miniaturized chips and the PC changed all that.

7.1.3 *The 1960s: Outer Space*

The idea that humans might be able to move above the surface of the earth has been a theme of imaginative writers for centuries. It was, however, both theory and experiment which enabled the Wright Brothers to attain controlled powered flight. The continuation of this line of device development has now yielded routine commercial jet aircraft travel at heights of ~ 10 km and speeds of ~ 800 km/hr with special purpose jets reaching heights of over 50 km and speeds in excess of 3000 km/hr. And then, in the 1950s, a quantum change in this aeronautical progression occurred.

In 1957 the world became mesmerized by the first artificial satellite — Sputnik I launched by Russia — a 80 kg 40 cm diameter sphere circling the earth every 90 minutes. It broadcast shortwave telemetry consisting of repeated *beeps* modulated with thermal information. The USA then accelerated its space program, placing increasingly heavier manned and unmanned satellites into orbit. The apex of this program occurred in 1969 when the USA accomplished the first human landing on the moon. This sequence of developments is encapsuled by a heterogeneous progression suggested in Fig. 7.1.

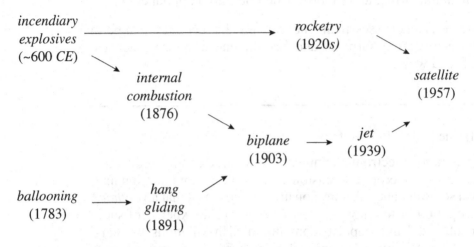

Fig. 7.1   Illustration showing the historical device progression leading to satellites.

Following the lunar landing program, space exploration expanded with three dominant foci:

(a) *Global Observation*
    Satellite communication emphasis was expanded to include global observation and eventually global positioning.
(b) *Deep Space*
    Launches of instrumental payloads designed specifically to obtain telemetric data about the physical characteristics of planetary satellites.
(c) *Space Station*
    The intent of this program is to establish a suitably placed and equipped platform to serve as a base for research and extended manned space missions.

It is interesting to note that some of the devices of the immediately preceding decades — nuclear power and computers — have been indispensable to space explorations; the former has provided long-life radioisotope power sources for a range of instrumental packages and the latter has enabled extensive calculations and trajectory simulations.

Some Remarkable Space-Age Pioneers

Konstantin Tsiolkovsky was a self-taught Russian school teacher who in the late 1890s began writing some exceptional papers on science, rocketry, artificial satellites, and space stations; his work was generally ignored until the 1930s. Robert Goddard, an American physicist, experimented with rockets and various instrumentation packages beginning in the 1920s; only the Weather Office was mildly interested but by the early 1940s government research funds became increasingly available. Hermann Oberth, an Austrian medic, wrote a dissertation on rocketry — which was promptly rejected by the University authorities for being too speculative; however, a 1934 popular book version of it became a best seller and thereafter his work was supported by the Germany military.

7.1.4 *The 1970s: Diversity I*

Developmental activity in nuclear energy (~1940s), computers (~1950s), and outer space (~1960s) tended to overshadow the development of numerous other engineering activities during these decades. Nevertheless, considerable development occurred in agriculture, transportation, and domestic appliances. As a consequence of increased costs of petroleum fuels, many devices were also redesigned for energy conservation.

Further, biomedical engineering, which had its earliest beginnings already 400 years ago with lower-limb prosthetic devices, is deserving of special note. A particular stimulation occurred in the 1950s with the heart-lung machine and then with kidney dialysis devices and radioisotope pacemakers, both emerging in the 1960s. The 1970s also experienced the successful application of computer assisted tomographic scanning, expanded application of radioisotope tracers, and the development of improved bio-compatible materials for implantable devices. In the same decade, the laser became an operational device and found a range of vital applications — from land surveying to communication to eye surgery and many other applications. Additionally, various novel electronic sensors found numerous medical and industrial applications.

Finally, the 1970s also included three distinct cluster developments of particular relevance to engineering:

(a) *Transportation*
    The evolution of expanded global transportation required consider-able expansion and improvements of airports, seaports, and adjunct facilities.
(b) *Manufacture*
    Globally integrated manufacturing became established, involving also centralized coordination and strategic planning.
(c) *Energy*
    Several *energy shocks* appeared in the 1970s thereby highlighting the primacy of indigenous energy sources. International energy dis-tribution and monopolistic pricing became of intense interest, stimu-lating alternative energy-source technologies, including also energy conservation.

These three developments tended to emphasize the considerable over-lap between political-economic interests and engineering knowledge for its service functions; the terms *infrastructure* and *strategic development* became associated with the underlying organizational and functional aspects of these cluster developments.

Consumerism

By the end of WWII, engineers and applied scientists were widely recognized for their ability *to get things done*. Also, society had by then passed through several decades of restrictions and often hardship. With the post-war economy expanding considerably in the West and in East Asia, society developed a strong interest in consumer devices and engineers were now well able to meet this demand. Devices were then produced in quantity at acceptable costs — and hence purchased in quantity. The ensuing era of pro-moting the manufacture of goods and also protecting the consumer interest has given rise to the label the Age of Consumerism.

7.1.5 *The 1980s: Diversity II*

Following the high-visibility technological developments from the 1940s to the 1970s, the style and focus of engineering activity in the 1980s began to change. Of decreasing importance were the large scale and vast government sponsored highly focussed projects and in their place appeared a larger number of smaller and distributed engineering projects, many of which were expected to relate to regional interests. Dominant themes for engineering activity could nevertheless still be identified.

Among the subjects of engineering developmental interest to emerge in the 1980s were the various alternative energy conversions:

(a) Wind energy → electrical energy
(b) Solar energy → thermal or electrical energy
(c) Ocean tide/wave energy → electrical energy

In the area of transportation, several broadly based engineering developmental subject areas became of much interest:

(a) Electric cars
(b) Automotive fuel cells
(c) Catalytic converters for pollution abatement
(d) Enhanced energy efficiency processes and associated devices

Other related research, development, and demonstration activities involved the following:

(a) High speed trains
(b) Manufacturing versatility
(c) Cellular telephone communication
(d) Extended incorporation of computers in business, commerce, and the heavy industry
(e) Network security: electric power grids, transportation corridors, communication channels, hydrocarbon fuel supply, emergency/ service mobility, etc.

And again, as throughout the previous eras of engineering, a premium was placed on ingenious innovations.

7.2   Changing Functions of Engineering

The period ∼1940 to ∼1970 provided a remarkable demonstration of science and engineering in its contribution to the conduct of WWII and subsequent industrial, political, and economic development on a global scale. Social commentators and political leaders widely recognized the effectiveness of the application of advanced knowledge in addressing issues of national interest — and engineering responded with various initiatives.

7.2.1   *Scientific Model of Engineering*

Immediately after WWII, a broadly based consensus emerged among many engineering educators and leaders of professional and technical engineering associations that it was appropriate to expand the scientific focus in engineering education. Modern physics and chemistry together with advanced applied mathematics became expanded subjects in engineering curricula beginning in the 1950s and 1960s, displacing thereby traditional hands-on engineering subjects such as drafting, machining, and surveying — and these subjects tended to reappear in other post-secondary educational institutions such as Technical Schools and Applied Arts Community Colleges.

These curricula changes occurred at about the time as the rise of solid-state physics found a niche as electronics in departments of electrical engineering; similarly, physical and polymer chemistry became of increasing importance to departments of chemical engineering. Then, the appearance of mainframe computers on university campuses provided very substantial computational opportunities in the application of differential and integral calculus in all branches of engineering — demanding as a consequence an expanded focus on advanced computational mathematics. Indeed, the association of some engineering disciplines with applied science became so strong and visible that Applied Science often became synonymous with Engineering as a field of academic specialization.

Departmental and disciplinary changes also took place, reflecting the historical adaptation of engineering to scientific developments and changing societal interests. Departments of Industrial/Manufacturing Engineering emerged as did Departments of Aeronautics and Astronautics. New departmental foci such as Engineering Science, Nuclear Engineering, Biomedical Engineering, and Materials Engineering were established. And, once the computer mainframe-to-distributed-system transition had become evident, programs and eventually Departments of Computer (or Software) Engineering similarly appeared.

7.2.2 *Military-Industry Stimulation*

Circumstances of WWII led to the formation of strong but selective collaborations between military interests and industrial design/manufacturing capability. Governmental subsidization provided the initial stimulant and continuing lubricant for these alliances, thereby contributing to substantial growth for some corporations and considerable innovation in specialized engineering subject areas. This symbiosis remained in place following termination of WWII and continued to be an important component in the waging of the *Cold War*, ~1948–~1988.

From an engineering point of view, this military-industry alliance fostered several unique practices of engineering relevance:

(a) *Science-Engineering Integration*
    Scientists and engineers often worked jointly on projects, bringing a theoretical-physical view in direct contact with a design-manufacturing-operations perspective of devices. This introduced another feature of modern engineering development: rapidity of device development.

(b) *Rapidity of Device Development*
    For reasons of military priority and then civilian urgency, procedures were developed for highly interactive RDD (Research, Development, Design) activity quickly leading to prototype device demonstration. Engineers now became accustomed to a very dynamic and mobile style of engineering practice which became characterized by a new feature of systems of devices: complexity of systems.

(c) *Complexity of Systems*
    The increasing rapidity of technical development together with an increasing knowledge base, led to considerable diversity of component development for multiple or integrated purposes. Benefits of prior developments and production experience were then realized, leading to considerable operational diversity. A very large range of integrated devices for specialized applications could now be effectively introduced.

One interpretation of the above development was that device operation became more evidently and directly dependent upon other devices, somewhat akin to humans depending upon other humans for the attainment of specified goals. Component integration and system failure avoidance now became of increasing importance.

Another feature which emerged from the post-WWII effort to apply wartime production rapidity to civilian goods production, was a subtle reconsideration of patenting: patenting was often put aside in the interest of quick market entry and early market dominance. The reason for this is that patenting can be a slow process relative to time-to-market of some devices and may also require disclosure of some in-house proprietary information. Simultaneously, the notion of device innovation, proprietary knowledge, and trade secrets gained in importance; a critical question now related to device modifications for marketing enhancement without violating existing patents.

Stimulation of Invention/Innovation

It is frequently thought that under extreme pressure such as wartime, the symbiosis of military-industry activity may lead to an increasing frequency of inventions; the alternative view holds that invention proceeds at a pace unaffected by social, industrial, commercial, or ideological pressure. Experience has shown only that the speed of invention-to-application and the range of diversity-of-innovation can be greatly advanced by organizational practices common under regional or national pressure, but inventiveness — like organic life — tends to proceed by its own independent complex rhythm, including sporadic clusters of inventions.

7.2.3 *Global Industrial Expansion*

The post-WWII continuation of the Military-Industry alliance led to a number of high-tech firms gradually adopting a stance specifically directed to consumer goods in both the domestic and global market. Possession of a substantial in-house knowledge and R&D base, coupled with an enhanced device testing and production capability, suggested to many corporations that civilian goods and services could substitute for a particular military device manufacturing capability. Among the most appropriate technologies for the global commercial market were telecommunication, electronic audio-video systems, and medical service devices. At another level of market expansion into the global commercial arena, one could identify the aerospace, computers, nuclear, automotive, and the pharmaceutical industries.

However, global expansion proved to be highly uneven — competition, resource distribution, unreliable infrastructures, available labor skills, and regional political interests often proved to be unsettling. Nevertheless, many corporations established diverse global production and distribution niches.

For engineers, the marketing of manufactured products now introduced two changes, both involving fast long distance communication:

(a) Branch-plant device manufacture
(b) Partner-firm device compatibility

Then, to add to the demands on engineering practice, new design procedures and new criteria for devices and systems became increasingly critical:

(a) *Time-to-Design*
    Parallel — as distinct from sequential — design and testing procedures were instituted in order to reduce the time from concept-to-market.
(b) *Operational Reliability*
    Devices had to possess increasing reliability under a range of application and usage patterns.
(c) *Production Mobility*
    Product schedules and component streams had to be synergetically varied, often very rapidly by at-a-distance directives.
(d) *Component Compatibility*
    High priority had to be assigned to component manufacture in different regions and at different times, and subjected to specialized compatibility testing for final operational integration.

The latter part of the Modern Engineering period thus introduced both an international flavor and increasing attention to flexible manufacturing. Intellectual demands tended to increase as did the expansion of feedback in systems management.

Growth and Development

The terms *growth* and *development* have become increasingly common expressions during the Modern Engineering era, often used
with little thought to their meaning and distinction. Correctly used,
growth refers to a quantitative increase and *development* refers to a
qualitative improvement. Note therefore an important consequence
to the theory and practice of engineering: *growth* is ultimately
limited because of the finiteness of the human habitat whereas *development* is evidently unbounded because it relates to the human imagination. Engineering ingenuity is becoming increasingly recognized
as a significant contributor to *development*.

7.3 Public Apprehensions

The 1950 → 1965 part of the Modern Engineering period was characterized by an expanded interest in various consumer goods and services,
but the following 1965 → 1990 part of this engineering period became
substantially influenced by a number of social apprehensions and organized protest movements. While initially some of these concerns had an
anti-war focus and subsequently developed a social justice interest, there
emerged also an environmental and anti-technology emphasis.

To put these public apprehensions into perspective, it became important to recognize that engineered devices are not simply inert or neutral
assemblages of matter: devices often became important in an individual's
life style and enablers of community aspirations, both invariably involving changing personal practices; and these changing practices may well
imply some objectionable directions as well as subtle ideological adjustments, each contributing to a varying level of individual anxiety. The
frequently heard assertion that engineers only design and make devices
which are *neutral* in some fundamental sense and that it is society which
decides on its positive or negative applications, is both evasive and hollow. Engineers are increasingly expected to be well versed in the range
of trade-offs associated with new devices and be able to elaborate on
the linkages between a particular design and its range of functions and
impact.

Some of these apprehensions may relate to specifics of device origin,
device promotion, and device market entry. But a most crucial aspect

relates to perceived bounds of device function and its impact — that is, the emergence of some sense of the ultimate use and purpose of a device. These factors thus identify details of birth, market entry, and societal impact of a device, Fig. 7.2.

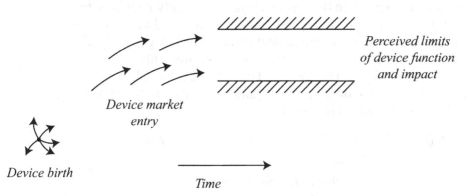

Fig. 7.2 Graphical depiction of device birth, market entry, and perceived bounds of device function and impact.

Some technological developments and industrial processes may be identified as particular stimulants to these societal apprehensions. In particular, two generic classes of device properties which relate to societal dissatisfaction have emerged as especially important in Modern times:

(a) *Device Operations*
Some devices evidently contribute to apprehensions by virtue of their operation: automobile pollution, jet-engine noise, paper mills which eject contaminating effluent into rivers and lakes, power plants which vent greenhouse gases and suspended particulates, smelting and metal processing plants which discharge poisonous heavy metals into open holding ponds, processing plants which produce carcinogenic and other toxic waste products, etc.

(b) *Device Failures*
Device failure may often lead to considerable harm and suffering: automobile tire blowout, train derailment, dam failure, aircraft crashes, fire attributable to faulty insulation, building and scaffolding collapse, bridge collapse because of material degeneration, chemical explosions attributable to failed sensors, toxic spills due to pipe bursting, poison gas releases due to containment wall ruptures, etc.

Societal apprehensions of the 1970s to 1990s associated with these and other engineered device failures and operational shortcomings have led to much public discussions, some legislation, and introspection by engineers. As a consequence, many engineers have specifically sought to develop a greater sensitivity to public safety and device risks, introducing also substantial changes in design, testing, construction, and operational methodologies of devices. Beginning in the 1970s a number of initiatives were introduced to expand on this engineering-and-society interface. For example, professional engineering associations introduced new sections to focus specifically on the encouragement of an engineering-society dialogue and some universities introduced new programs with an emphasis on science, technology, and society. Thus, the feedback linkages of (4.12) and (5.2) could be extended to recognize an active reciprocal feedback/feedforward dialogue suggested by

$$N(t) \;\rightarrow\; E(t) \;\rightarrow\; D(t) \;\rightarrow\; S(t) \;\rightarrow\; R(t)$$

$$\textit{feedback/feedforward} \atop \textit{dialogue}$$

(7.5)

as an addition to the engineering connectivity. Note also that this curved arrow suggests both intellectual dialogue and information flow while the straight arrows imply material flow, though they may also be accompanied by some implicit information about the materials (e.g. instruction on device assembly, usage, etc.).

Meaning of Technology

The word technology is widely used as a catch-all phrase though a widely accepted definition has yet to emerge. This term is often erroneously taken to imply the existence of some extraneously imposed force or power responsible for many of the perceived shortcomings of contemporary life: atmospheric contamination, traffic congestion, loss of privacy, depression, hectic life-styles, etc. A more informing view suggests that technology represents the planning, making, implementing, and using devices for the attainment of specific goals. Hence, technology is not some *one thing* but an extended *means to attain particular ends* and about which individuals may, for a variety of reasons, possess uncertain and ambiguous feelings. Evidently, engineers are major — but not the only — contributors to technology.

7.4   Modern Engineering: Expanded Reach of Devices

Consumer interest for rapid device production during the Modern Engineering period (1940–1990) added considerably not only to analysis and design recursion loop first introduced during the Renascent Engineering period but also to device manufacture and implementation:

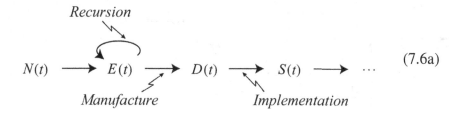

$$ \tag{7.6a} $$

Additionally, repository accumulations discussed in the immediately preceding Expansive Engineering period (1850 $\to$ 1940), also expanded

$$ \cdots \longrightarrow S(t) \searrow R(t) \tag{7.6b} $$

Increase

and this repository $R(t)$ became further exacerbated by the increasing scale of production. Dominant among these waste accumulation increases were the following sources:

(a) Toxic tailings from resource extraction
(b) Poisonous heavy metals from metal processing
(c) Carcinogenic substances from high-grade material production
(d) Radioactive materials from the nuclear industry
(e) Greenhouse gases and suspended atmospheric contaminant production from expanded power generation and transportation

All of these processes can be characterized as waste productions resulting in contaminant flows. While some can be separated and retained by sequestering, ultimately these materials add to accumulation in the repository $R(t)$ of the engineering connectivity progression; the fractions which are not sequestered constitute contamination of the atmosphere, water, and soil and hence of nature. These flows are accounted for by an extension to (6.18) to yield the connectivity

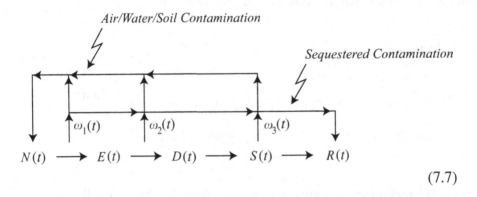

$$(7.7)$$

Here $\omega_1(t)$ identifies the waste flow component from natural material extraction and primary processing, $\omega_2(t)$ as the waste flow component from secondary processing and device production, and $\omega_3(t)$ is the waste flow component from device utilization. Parts of each of these waste streams appear as suspensions in the atmosphere, as diffused components in water, or as concentrations in soil; thus they contribute to an unrecoverable depositional material contaminant which constitutes a return flow to nature. The sequestered components appear in the terrestrial repositories $R(t)$, first introduced in (6.19).

We emphasize that there exists however a component contribution to $R(t)$ accumulations which often poses a particular dilemma since it relates to diverse individual and community choices: $S(t) \rightarrow R(t)$ also involves used appliances, worn tires, dated newsprint, variety of food packaging (cartons, bottles, cans, plastic wrapping), obsolete computer components, failed tools and utensils, etc.; indeed, devices are often discarded for reasons of fashion or prestige. This expanded waste material flow increased considerably during the 1940 $\rightarrow$ 1990 period and has prompted many commentators to characterize this time interval as the *throw away* or *built-in obsolescence* era.

The gradual saturation of repository storage $R(t)$ capacity, largely involving municipal dumps and landfill sites as well as industrial scrap and soil contamination areas, has become a significant problem for many communities and corporations. One mitigation process which emerged with some prominence during the middle part of the Modern era was the notion of recycling and thereby involving engineers in a previously little-practised activity suggested by an addition to the engineering connectivity

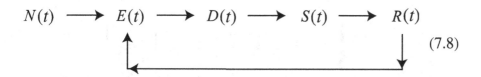

$$N(t) \longrightarrow E(t) \longrightarrow D(t) \longrightarrow S(t) \longrightarrow R(t) \tag{7.8}$$

In effect, some $R(t)$ accumulations are redirected to become part of the original device production process but first requiring device disassembly and material separation. For glass, metallic cans, and newsprint, this involves separation of lids, labels, and ink and is readily accomplished providing batch sizes are appropriate and inflow rates are matched to processes. For junked automobiles and other transportation devices (aircraft, buses, railcars), material stream separations may — depending upon the purity required — be labor intensive unless the original design takes into account eventual disassembly; reconditioning of components such as engine blocks, starters, alternators, etc., has also become a successful recycling activity. For devices such as waste tires, obsolete computers, and various polymers (e.g. PCBs, styrofoam, etc.), significant problems of disposal remain.

Combining (7.5) to (7.8) together with that of (6.20) now characterizes the connectivity of Modern Engineering as follows:

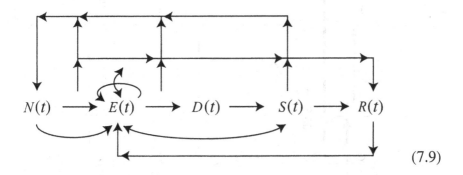

$$N(t) \longrightarrow E(t) \longrightarrow D(t) \longrightarrow S(t) \longrightarrow R(t) \tag{7.9}$$

Note that this depiction illustrates the trend of engineering theory and practice toward material closure.

Table 7.1 illustrates the continuing progression of engineering now including Modern Engineering ($\sim$1940 $\rightarrow$ $\sim$1990).

Table 7.1   Addition of Modern Engineering (~1940 → ~1990) to the evolution of engineering.

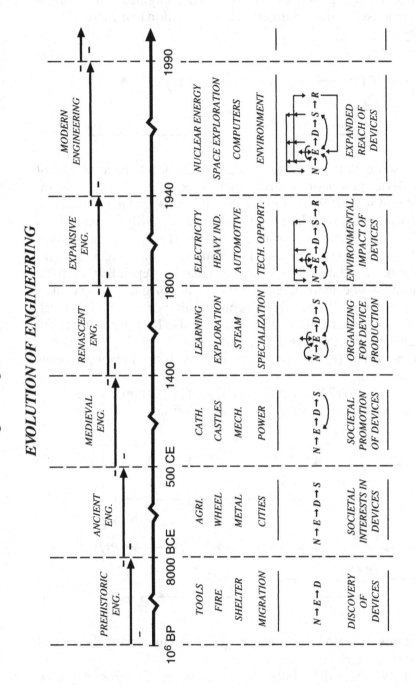

7.5 To Think About

- Linkages such as $A(t) \rightarrow B(t)$ invariably imply matter and energy transformations unavoidably resulting in a variety of by-products. List such transformations for the electric energy production industry, the transportation industry, and the electronic industry. Include also process inefficiencies.
- Wartime has proven to be a remarkable stimulus in establishing support for engineering activity and increasing rates of device production. Itemize some pros and cons to this hurried process with respect to rational device development, production, and deployment.
- Describe processes and methods which can be used to mitigate and reduce the waste streams identified by $\omega_1(t)$, $\omega_2(t)$ and $\omega_3(t)$, as suggested in (7.7).

CHAPTER 8

Contemporary Engineering

(~1990 → ~2000+)

Prospects for Closure

8.1 Engineering of the Present

The period of engineering interest in this chapter is taken to have begun about ten years ago and, in an approximation, extends to the near future of perhaps five to ten years. An examination of evolutionary changes in engineering over such a short and contemporary time interval — and which may relate meaningfully to the near future — presents many uncertainties because the intrinsic developmental noise may mask or distort important emerging patterns. For this reason and as an effective starting point, we choose to be guided by the recent and evident changes associated with the several influences directly affecting engineering $E(t)$ of (7.9); that is, the relations of interest is given by the process content of the dash-enclosed rectangle of Fig. 8.1.

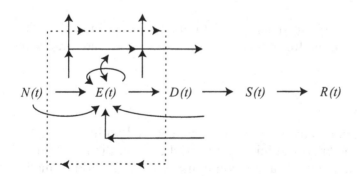

Fig. 8.1   Factors affecting Contemporary Engineering theory and practice.

Our objective is now to examine each of the indicated influences suggested by the arrows in this figure:

(a) $N(t) \xrightarrow{\quad} E(t)$

Fractional changes of most non-renewable resources of importance to engineering have been modest with two important exceptions: liquid-hydrocarbon petroleums continue to be locally depleted at a significant rate; mining, extraction, and processing continue to contribute to toxic and non-toxic waste production and accumulation — though at a reduced rate and more of it now in sequestered form.

(b) $E(t)$

Engineering educational programs and design recursion have experienced continuing development; to be noted is an expanded emphasis on device reliability and adaptability, mathematical simulation analysis, sensitivity to public preferences, and expanded consideration of environmental impact.

(c) $E(t)$

The Firm-Engineer linkage has not experienced major change in recent years though the continuing evolution of globalization has been of considerable importance to changing corporate foci of relevance to the practice of engineering.

(d) $N(t) \qquad E(t)$

Engineering has retained its foundational focus on the scientific method, as a basis for its continuing theoretical and practical development.

(e) $E(t) \xrightarrow{\quad} D(t)$

Design and manufacture of devices have evolved with an increasing emphasis on energy efficiency, material conservation, reduction of waste streams, flexible and reconfigurable manufacturing, small-batch niche production, expanded supplier-manufacturer-distributor integration, and improved devices for sequestering environmental pollutants.

(f) $E(t) \longrightarrow \cdots \longrightarrow S(t)$

The influence of an evolving societal interest on engineering continues to be significant — now sometimes expressed as involving
social control. Especially to be noted are the engineering responses
to broadly-based interest groups and political attempts at mediation;
these have led to increasing engineering designs toward more user-
friendly and more environmentally compatible devices.

(g) $E(t) \longrightarrow \cdots \longrightarrow R(t)$

Recycling of homogeneous devices has continued to increase. Complex heterogeneous devices (e.g. household appliances and automobiles) are now often designed for efficient disassembly and
material stream separation, but others (e.g. PCs, certain electronic
devices, etc.) continue to pose problems sufficiently so that some
jurisdictions require manufacturers to *take back* their obsolete
devices. The expression *back-end-of-material-flow* has been introduced as a descriptive term for this material accounting process. For
the first time, material closure is commonly expressed as a desirable
goal.

Of the above listings, it is evident that engineering has undergone pronounced changes during the Contemporary period and these are expected
to continue.

The Evolution of Engineering

The generally increasing and continually changing engineering connectivity, as illustrated at the end of each of Chapters 2 to 7 and
summarized in Fig. 8.1, illustrates a distinct and pervasive feature of engineering: it evolves with time in accordance with externally imposed and internally generated influences — and is thereby
becoming a *global adaptive evolutionary system*.

8.2 Continuing Device Evolution

Devices come about as a result of engineering thought and skill, and their
functions in a societal milieu are affected by personal choices and group

preferences. This societal milieu is a complex amalgam of individual and community aspirations, beliefs, and ideologies which by feedback provide preference indications to engineers for device development and manufacture. These various interests are then integrated with industrial and commercial considerations, eventually yielding successive generations of devices — including also decline of some existing device species. Collectively one can identify four changing features of device evolutions which relate in a special way to Contemporary times: diversity, emergence, decline, and uncertainty.

8.2.1 *Continuing Device Diversity*

Diversity is a most pronounced feature of device evolution. The following examples suggest such continuing evolutions in Contemporary times:

$$
\cdots \to \textit{Telephony} \to \cdots \to
\begin{array}{l}
\nearrow \quad \textit{performance enhancement} \\
\nearrow \quad \textit{cellular expansion} \\
\quad \textit{fiber-optics substitution} \\
\searrow \quad \textit{information storage capability} \\
\searrow \quad \textit{expanded integration} \\
\quad \cdots
\end{array}
\qquad (8.1a)
$$

$$
\cdots \to \textit{Computers} \to \cdots \to
\begin{array}{l}
\nearrow \quad \textit{speed/memory increase} \\
\nearrow \quad \textit{size/weight reduction} \\
\quad \textit{enhanced encryption} \\
\searrow \quad \textit{diverse applications} \\
\searrow \quad \textit{operational choices} \\
\quad \cdots
\end{array}
\qquad (8.1b)
$$

$$
\cdots \to \begin{array}{c} \textit{Energy} \\ \textit{Production} \end{array} \to \cdots \to
\begin{array}{l}
\nearrow \quad \textit{polluting by-product abatement} \\
\nearrow \quad \textit{solar-electric conversion} \\
\quad \textit{environmental power sources} \\
\searrow \quad \textit{minipower distributed systems} \\
\searrow \quad \textit{electricity-heat cogeneration} \\
\quad \cdots
\end{array}
\qquad (8.1c)
$$

Considerable interest exists in these engineering areas and a substantial corporate investment continues to be sustained.

8.2.2 *Threshold Device Emergence*

A number of generic devices and related areas of activities seem to hold promise for significant development in the near and intermediate term:

$$\cdots \to \begin{array}{c} \textit{Automotive} \\ \textit{Engines} \end{array} \to \cdots \to \begin{array}{l} \nearrow \quad \textit{fuel-cell substitution} \\ \textit{hybrid implementation} \\ \searrow \quad \cdots \end{array} \qquad (8.2a)$$

$$\cdots \to \textit{Bioengineering} \to \cdots \to \begin{array}{l} \nearrow \quad \textit{expanded prosthesis} \\ \textit{genetic engineering} \\ \searrow \quad \cdots \end{array} \qquad (8.2b)$$

$$\cdots \to \begin{array}{c} \textit{Space} \\ \textit{Exploration} \end{array} \to \cdots \to \begin{array}{l} \nearrow \quad \textit{deep space missions} \\ \textit{atmospheric-terrestrial diagnosis} \\ \searrow \quad \cdots \end{array}$$

$$\qquad (8.2c)$$

$$\cdots \to \begin{array}{c} \textit{Water} \\ \textit{Resources} \end{array} \to \cdots \to \begin{array}{l} \nearrow \quad \textit{hydroponics} \\ \textit{purification/distribution} \\ \searrow \quad \cdots \end{array} \qquad (8.2d)$$

$$\cdots \to \begin{array}{c} \textit{Environmental} \\ \textit{Sensors} \end{array} \to \cdots \to \begin{array}{l} \nearrow \quad \textit{effluent composition} \\ \textit{air quality} \\ \searrow \quad \cdots \end{array} \qquad (8.2e)$$

$$\cdots \to \begin{array}{c} \textit{Pollution} \\ \textit{Abatement} \end{array} \to \cdots \to \begin{array}{l} \nearrow \quad \textit{various filters} \\ \textit{material flow control} \\ \searrow \quad \cdots \end{array} \qquad (8.2f)$$

$$\cdots \to \begin{array}{c} \textit{Nano} \\ \textit{Technology} \end{array} \to \cdots \to \begin{array}{l} \nearrow \quad \textit{atom-scale construction} \\ \textit{molecular computation} \\ \searrow \quad \cdots \end{array} \qquad (8.2g)$$

By many indications, these areas hold much research, development, design, and manufacturing promise and are often well supplied with funding support.

8.2.3 *Apparent Device Decline*

The last decade has seen a decline in some devices and associated manu-facturing in the corresponding industries. Large-scale mining and heavy shipbuilding appear to be particularly affected.

8.2.4 *Continuing Device Uncertainties*

For a number of reasons, several devices and associated technologies face uncertainties; among these, two examples appear to be railroading and commuter transport.

Thus, like biological species, human-made devices also possess some evolutionary features: they are born, they might mature and prosper, they might hold steady, they might transmute, they might experience accidental death, they might decline, but — and this is significant — they may never totally vanish. Additionally, their life cycle is much dependent upon the technical-societal environment for their existence.

Evolution of Devices

Devices may often be viewed as extended human prostheses: the hammer as an extension of the hand, the telescope as amplifiers of the human eye, etc. Devices may also be viewed as providers of important commodities: clean water by filters, warmth by heaters, motive power by electric generators, etc. A more general and also deeper view suggests that devices evolve with time to provide *ends* as well as *means* and they achieve these *ends* and *means* by *birth* in design, manufacture, and implementation; subsequently they *fulfill* some societal expectations, and finally they serve as stimulants to a next *generation* of devices.

8.3 Shifting Design Criteria

It is frequently stated that a characteristic of engineering is that it often seeks technical solutions to technical problems and these solutions have invariably been attained on the basis of narrow design criteria involv-ing material composition, geometrical configuration, and economic justification. For the primal engineering progression, one may list these considerations as part of engineering thought and actions now extended to include the recursive design loop

$$N(t) \to E(t) \left\{ \begin{pmatrix} thought \\ actions \end{pmatrix} : \to \begin{pmatrix} material\ composition \\ geometric\ configuration \\ economic\ Justification \end{pmatrix} \right\} \to D(t) \cdots \tag{8.3a}$$

to suggest components of a rational process as a variation of (3.9).

Contemporary considerations which relate to an evolving societal interest now demand the inclusion of broader design considerations such as user psychology, community ideology, and environmental impact. Contemporary engineering has thus further evolved as

$$N(t) \to E(t) \left\{ \begin{pmatrix} thought \\ actions \end{pmatrix} : \to \begin{pmatrix} material \\ geometric \\ economic \end{pmatrix} \to \begin{pmatrix} psychology \\ ideology \\ environment \end{pmatrix} \right\} \to D(t) \to S(t)$$

$$\tag{8.3b}$$

where, for reasons of notational clarity, all recursive loops are dropped. Thus an expanded perspective is evident.

It is important to emphasize that while additional engineering thought and actions leading to device design have emerged as especially acute during Contemporary times, elements of these criteria have been noticeable much earlier. It appears that the beginnings of these expanded considerations can be traced to the Renascent Engineering era as practices desirable to a societal interest based on emerging considerations of preservation of nature, the idea of conservation of resources, and a recognition that both preservation and conservation could be more effectively realized by extended consultations.

8.3.1 *Preservation of Nature*

The birth of the factory system and large scale smelting — emerging some 250 years ago in Europe — had established the recognition that the making of some devices may involve considerable despoilage and devastation of the landscape. Indeed, European royalty and aristocrats had already then found it appealing to set aside large tracts of virgin land which were to be free of industrial activity; the motivation was evidently their interest in hunting game as a sport mixed also with the desire to display their wealth and suggest influence.

By the late 1800s, a popular movement arose in North America promoting the establishment of large national parks in order to preserve the flora and fauna for the enjoyment of society; National Parks such as Yellowstone, Yosemite, and Banff, were among the first such preserves to be established and many more have since been added. Preservation has continued to be a dominant management objective for these major resources.

8.3.2 *Conservation of Resources*

The idea of preserving nature in the form of National Parks proved to be most successful in soliciting widespread societal approval. By the late 1900s, it became generally accepted that this notion should be extended to the conservation of natural resources in general. Three specific interests were identified:

(a) *Resource Depletion*
    Some non-renewable natural resources were becoming locally depleted and hence increasingly costly; this led to problems of risky exploration, distant transportation, and international negotiations.
(b) *Environmental Contamination*
    Extraction and processing of mineral resources for purposes of device production became associated with by-products of critical adverse environmental impact in its effects on the quality of the living habitat.
(c) *Waste Accumulation*
    Wastes from device manufacture as well as from spent device accumulations were often posing significant hazards, demanding therefore expanded sequestering and confinement.

These three features of relevance to resource conservation appear as specific components of the engineering connectivity

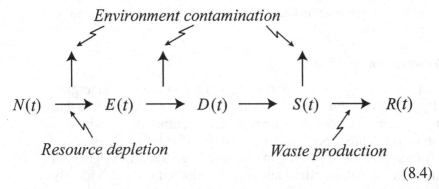

$$\text{(8.4)}$$

A societal belief emerged which asserted that resolution — or at least abatement — of this aspiration had to be approached collectively and that this had to involve expanded public consultation.

8.3.3 *Legislated Public Consultation*

Public consultation processes were variously viewed as consisting of three formal kinds of activities:

(a) *Impact Assessment*
   Specific engineering projects were subjected to a broadly based legislatively mandated impact assessment, focusing in particular on relevant technology, environmental impact, and potential community life disruption.
(b) *Citizen Advisory*
   Citizen advisory committees were established by *order-in-council* in order to provide views and advice on emerging society-related issues relative to specific engineering project development.
(c) *Public Hearing*
   Major engineering projects were subjected to legislated public hearings in which credible non-profit interveners were provided with financial resources for relevant research and presentation costs.

These three considerations identify further communication and assessment processes as part of the engineering connectivity as an extension of (8.4):

$$
\begin{array}{c}
\textit{Impact assessment} \\[2pt]
N(t) \longrightarrow E(t) \longrightarrow D(t) \longrightarrow S(t) \longrightarrow R(t) \\[2pt]
\textit{Citizen advising and} \\
\textit{public hearings}
\end{array}
\qquad (8.5)
$$

Note that in all these processes — assessment, advising, and hearings — engineers and the public exchange information, test assumptions, and seek an accommodation.

And so the workings of engineers — traditionally based only on material, geometrical, and economic considerations — expanded in contemporary times to additionally involve psychological, sociological, ideological, as well as philosophical and anthropological considerations; engineers are now of necessity more inclusive in outlook and their profession is becoming increasingly connected to a wider world.

Foundations of the Engineering Progression

The evolution of engineering in this text is characterized by a progressive linkage connectivity integrating the nodes $N(t)$, $E(t)$, $D(t)$, $S(t)$, and $R(t)$. At their most foundational level, these nodes constitute a linkage between configurations of matter and energy with sources of knowledge and ingenuity:

$$\text{Sources of knowledge and ingenuity}$$

$$N(t) \longrightarrow E(t) \longrightarrow D(t) \longrightarrow S(t) \longrightarrow R(t) \qquad (8.6)$$

$$\text{Configurations of matter and energy}$$

Engineers are the critical contributors to the meshing of knowledge and ingenuity with a range of configurations of matter and energy for the purpose of meeting the economic, political, and religious interests of society.

8.4 Emerging Foci

In addition to shifting device design criteria, Contemporary times also point to the emergence of other foci for engineering practice. It has now become essential for engineers to expand their perspective and be alert to changing methodologies especially as they relate to advanced modeling and impact assessment. For example, device design now requires a focus on *cradle-to-grave* material accounting, energy consumption optimality, and higher operational reliability.

A birds-eye-view of Contemporary Engineering suggests the emergence of three particular foci for engineering theory and practice: the development of devices with improved *matter-energy efficiency*, the expansion of *industrial ecology* systems, and the primacy of complex *development* as distinct from simple *growth*.

8.4.1 *Matter-Energy Efficiency*

The history of science and engineering illustrates some remarkable cases of devices which, while reducing matter and energy requirements, have led to most pronounced increases in device effectiveness. Numerous examples of development of devices which accomplish *more-with-less* can be cited.

First practical jet engines (~1940s) averaged about 1 g of engine weight to produce 1 W of power; by the mid-1950s, this weight requirement for the same power was reduced ten-fold. The initial energy cost of synthesizing NH_3 (ammonia for fertilizer production) in the 1930s involved ~500 MJ/kg and its current requirements are less than 50 MJ/kg. These remarkable ten-fold efficiency improvements with time appear as details of the primal progressions:

$$N(t) \rightarrow E(t) \rightarrow D(t)\{ jet\ engine\ power/weight,\ W/kg \nearrow \}, \quad (8.7a)$$

$$N(t) \rightarrow E(t) \rightarrow D(t)\{ fertilizer\ energy\ cost,\ J/kg \searrow \}. \quad (8.7b)$$

Also, first computers of the 1940s filled large rooms, required significant electrical power, needed extensive air conditioning, involved high maintenance costs — and still were unreliable in operation. The reason for size, power, and inefficiency were due to the extended copper-wire circuitry and the very large number of vacuum triodes. Then, in 1948 the transistor was invented, followed by the integrated circuit in 1958 and the microprocessor in 1970. It is this combination of devices which enabled computers to become compact, fast, reliable, less costly, and requiring less matter to build and less energy to operate. This remarkable ingenuity progression is suggested by

$$N(t) \rightarrow E(t) \begin{Bmatrix} miniature & reduced & high\ speed \\ circuit & \rightarrow heat & \rightarrow reliable \\ components & dissipation & switching \end{Bmatrix} \rightarrow D(t) \begin{Bmatrix} small,\ fast, \\ effective, \\ computers \end{Bmatrix}.$$
$$(8.8a)$$

By any measure, orders of magnitude of improved operation have been attained while simultaneously requiring less matter and less energy.

A similar ingenuity progression can be cited in the automotive industry:

$$N(t) \rightarrow E(t) \begin{Bmatrix} lighter & reduced & reduced \\ cars & \rightarrow fuel & \rightarrow atmospheric \\ & consumption & pollution \end{Bmatrix} \rightarrow D(t) \begin{Bmatrix} more \\ efficient \\ transportation \end{Bmatrix}.$$
$$(8.8b)$$

Often a progression component may be isolated for its multiple and integrated choices. A case in point is the development of light-emitting diodes and liquid-crystal devices for purposes of illumination. In applications such as traffic lights and instrumentation panels, one may well identify overlapping developments:

$$N(t) \rightarrow E(t) \left\{ \begin{array}{l} \textit{suitable} \\ \textit{electronic} \\ \textit{junctions} \end{array} \begin{array}{l} \nearrow \textit{ reduced energy} \\ \nearrow \textit{ increased longevity} \\ \searrow \textit{ faster response} \\ \searrow \textit{ smaller size} \end{array} \cdots \right\} \rightarrow D(t).$$

(8.8c)

Such consequent changes in engineering theory and practice are increasingly appearing in Contemporary times suggesting much opportunity for devices which accomplish *more-with-less*.

8.4.2 *Industrial Ecology*

Industrial ecology is a concept which takes its cue from the complex web of food chains existing in nature. The underlying pattern is the recognition that a factory uses natural resources and in the process of manufacturing devices also generates some wastes:

$$\begin{array}{c} N(t) \rightarrow Factory \rightarrow Device \\ \downarrow \\ Waste \end{array}$$

(8.9)

The premise is that one may identify some relevant first tier factory operation with a waste output which could constitute an essential input to some second tier factory system, Fig. 8.2.

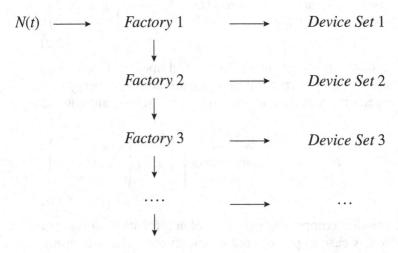

Fig. 8.2   Graphical depiction of an industrial ecology network.

Typically, Factory 1 represents some fossil burning electricity producing plant — in which case Tier 1 could more specifically be written as $N(t) \rightarrow$ (energy conversion factory) $\rightarrow$ (electric power) and its waste of heated water normally discharged into an adjacent river or lake. The second Tier factory could, for example, be a greenhouse which uses the waste heat from Tier 1 in its production of plants and vegetables; we add that such co-generation of electricity and heated water has long been installed in some regions for both domestic and commercial purposes.

One may note that a coal-burning power plant has other wastes of potential utility. For example, fly ash may be extracted and used for cement making. Then, depending upon the source of coal, its wastes may also contain useful quantities of rare-earth metals for the electronics industry and gypsum for building materials. Conventional factories may thus constitute serial or parallel lower tier installations.

An interesting mutualism can also be identified. Fish hatcheries also benefit from the power plant waste heat and the sludge from the fish tanks can serve as fertilizers for associated greenhouses. This suggest parts of an expanding network

$$
\begin{array}{l}
\qquad \downarrow \\
Factory\ i\ \rightarrow \quad \cdots \\
\qquad \downarrow \quad \searrow \\
Factory\ j\ \rightarrow\ Factory\ k\ \rightarrow\ \cdots \\
\qquad \downarrow \quad \searrow \quad \downarrow \quad \searrow
\end{array}
\qquad (8.10)
$$

designed for overall waste reduction and energy conservation.

In general, industrial ecology offers considerable efficiency appeal. Note however that the various tiers become coupled, requiring therefore high reliability of every higher order tier, and once a factory becomes so connected there will be restrictions on changing product lines and production rates. Hence, considerable planning and design needs to enter in the continuing trend towards industrial ecology.

8.4.3   *On the Primacy of Development*

The important distinction between growth and development has been noted in Chapter 7: *growth* as a quantitative increase and *development*

as a qualitative improvement. This focus on qualitative improvement —
rather than simply growth — has entered the consciousness of society and
of engineers and relates most specifically to device design, manufacture,
and operation. Such a specific and more intense focus — sometimes
associated with the concept of *creative economy* — appears to be an
outgrowth of increasing societal expectation and professional evolution;
evidently this emphasis is also a consequence of the increasing realization
that devices possess the quality of permeating human habits and community practices. Indeed, contemporary engineering itself constitutes a
development in time.

Contemporary Engineering: A Complex Adaptive System

The cumulative effect of a million year history on the expanding
engineering connectivity suggests the following as a compact summary characterization of Contemporary Engineering:

$$E(t) = E(t) \left\{ \binom{thought}{actions} : \begin{pmatrix} recognition\ of\ intricate\ processes\ in\ N(t) \\ capacity\ for\ development\ of\ D(t) \\ responsiveness\ to\ nonlinear\ feedback\ from\ S(t) \\ accommodation\ of\ boundedness\ of\ R(t) \\ ability\ for\ self\text{-}specialization\ within\ E(t) \end{pmatrix} \right\}$$

$$(8.11)$$

These five attributes thus establish Contemporary Engineering as a
complex adaptive system.

8.5   Contemporary Engineering: Prospects for Closure

Our continuing use of the evolving engineering connectivity (7.9) and
antecedent depiction of the various engineering eras, can now be concluded by alluding to the prospect of closure: that is, the flow of matter
approaches a heterogeneous flow pattern capable — in principle — of
long term sustainability. This closure can be depicted as the continuing evolution of engineering to define the Contemporary Engineering

connectivity:

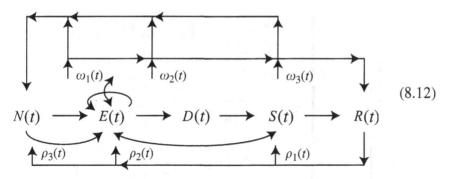

$$(8.12)$$

Here, the $\rho_i(t)$ symbols represent exit flow components from $R(t)$ associated with the functions of $\rho_1(t)$ = reuse, $\rho_2(t)$ = recycle, $\rho_3(t)$ = return to nature; the other symbols are used as previously defined. Recall, again, that the straight arrows primarily represent matter and energy flow while the curved arrows constitute information and ingenuity flow.

The engineering connectivity network (8.12) can be interpreted as a heterogeneous circuit-like flow diagram containing time dependent component processes $\omega_1, \omega_2, \omega_3, \rho_1, \rho_2, \rho_3$ and time dependent evolutionary and interactive progressive features associated with $N(t)$, $E(t)$, $D(t)$, $S(t)$, and $R(t)$. A nominal expectation is for $N(t)$ to be an exceptionally slow variable of time with the waste repository $R(t)$ declining from its present accumulation. The terms $E(t)$, $D(t)$, and $S(t)$ are expected to be continuing developmental variable of time and hence the process variables ω_i and ρ_i need to be adaptive functions of time.

Table 8.1 now completes our summary characteristics of the evolution of engineering.

8.6 To Think About

- Identify some contemporary highly controversial technological developments and discuss their features. Be specific in identifying the most critical economic, political, and religious issues.
- List contemporary examples of successful experiences with reuse and recycle; focus on especially acute devices.
- Explore engineering as it relates to the contemporary term *Sustainable Development*; include a reasonable definition of this term.

Table 8.1 Evolution of engineering now including the addition of Contemporary Engineering, 1990 →.

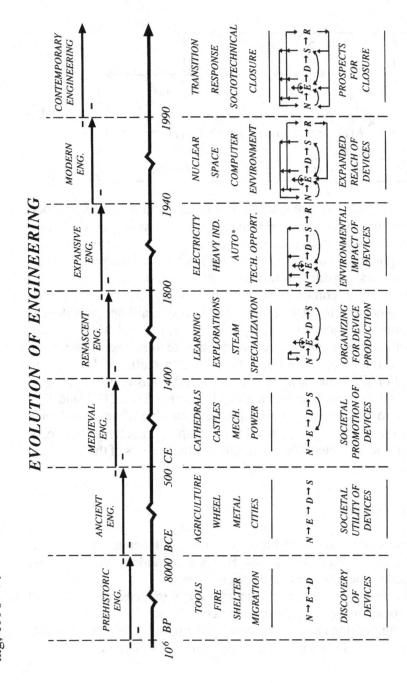

EVOLUTION OF ENGINEERING

	PREHISTORIC ENG.	ANCIENT ENG.	MEDIEVAL ENG.	RENASCENT ENG.	EXPANSIVE ENG.	MODERN ENG.	CONTEMPORARY ENGINEERING
	10^6 BP	8000 BCE	500 CE	1400	1800	1940	1990
	TOOLS FIRE SHELTER MIGRATION	AGRICULTURE WHEEL METAL CITIES	CATHEDRALS CASTLES MECH. POWER	LEARNING EXPLORATIONS STEAM SPECIALIZATION	ELECTRICITY HEAVY IND. AUTO° TECH. OPPORT.	NUCLEAR SPACE COMPUTER ENVIRONMENT	TRANSITION RESPONSE SOCIOTECHNICAL CLOSURE
	$N \rightarrow E \rightarrow D$	$N \rightarrow E \rightarrow D \rightarrow S$	$N \rightarrow E \rightarrow D \rightarrow S$	$N \rightarrow E \rightarrow D \rightarrow S$	$N \rightarrow E \rightarrow D \rightarrow S \rightarrow R$	$N \rightarrow E \rightarrow D \rightarrow S \rightarrow R$	$N \rightarrow E \rightarrow D \rightarrow S \rightarrow R$
	DISCOVERY OF DEVICES	SOCIETAL UTILITY OF DEVICES	SOCIETAL PROMOTION OF DEVICES	ORGANIZING FOR DEVICE PRODUCTION	ENVIRONMENTAL IMPACT OF DEVICES	EXPANDED REACH OF DEVICES	PROSPECTS FOR CLOSURE

PART C

Contemporary Context of Engineering

The progression $N(t) \rightarrow E(t) \rightarrow D(t) \rightarrow S(t) \rightarrow R(t)$ represents the core connectivity of engineering. Variations in this connectivity are stimulated by imaginative, pragmatic, and idealistic initiatives from $E(t)$ and $S(t)$ establishing thereby the distinguishing features of the contemporary context of engineering.

CHAPTER 9

Nature: Emergence and Implications

$$N(t) \rightarrow E(t) \rightarrow \cdots$$

9.1 Nature: $N(t)$

The making of ingenious devices is foundationally based on natural materials and phenomena. Attempts to seek an understanding of this physical reality has attracted much thought over the millennia.

One of the early influential philosophers who thought deeply about nature and its manifestations was Aristotle (384–322 BCE) of Greece. He accepted the ancient notion that nature could be viewed as consisting of four essences: earth, water, air, and fire. He then reasoned that by a selected superposition of the experiential qualities of cold, wet, hot, and dry, one could obtain the known properties of matter. A graphical depiction of this Aristotelian view of nature is suggested in Fig. 9.1.

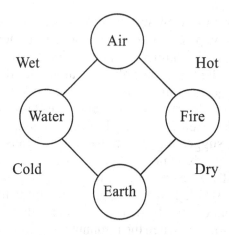

Fig. 9.1   The Aristotelian view of nature.

Though intuitively and geometrically appealing — and indeed this representation was taken as dogma for about 1500 years — it had to yield to the emergence of modern scientific thought generally taken to have its gradual beginnings in the mid 1500s with the works of Copernicus and subsequently many other inquisitive and creative minds.

With time and the dominance of a reductionist perspective, the objective features of nature eventually became interpreted in terms of the complex properties and interactions of nuclear, atomic, molecular, and biotic entities together with the subtle and powerful notion of action-at-a-distance and of matter-energy equivalence. This process of inquiry proved to be an exceptional accomplishment of the human intellect, yielding a remarkable progression of understanding of nature — and simultaneously providing a basis for the continuing evolution of engineering.

9.2　Nucleons and Atoms: Astronomical Basis

By the early 1900s the making of optical devices had become well established and soon proved to be of critical importance in suggesting a physical process for the origin of our Universe. In particular, optical telescopes provided images of distant stellar objects and optical emission spectroscopes allowed their compositional characterization. Interestingly, these two types of engineered devices relate to the world of the very large (i.e. astronomy) and the world of the very small (i.e. atomic structure), requiring especially exacting engineering design and precision manufacturing.

9.2.1　*Big Bang*

As late as the 1920s, the common view was that the Universe consisted of only the Milky Way Galaxy, possessing an enormous size then taken to be some 300,000 light years in diameter. Then, in the 1930s, Edwin Hubble (1889–1953), USA, identified variable stars at greater distances and found that these faint objects emitted characteristic electron orbital transition spectral lines at wavelengths which increase in proportion to their distances. This so-called Red Shift could be interpreted in terms of cosmological expansion suggesting that these very distant objects were mutually receding at speeds which increased with distance. Calculating backward in time, it was eventually concluded that between 10 and 15 billion years ago, a so-called Big Bang phenomenon could have occurred as the initiating event of our physical universe. Further supportive evidence was discovered in 1965 when the remnant radiation from this primal-source event was shown to be isotropically permeating all space.

9.2.2 Early Universe

A conceptual reconstruction of the Big Bang suggests a spontaneous singularity explosion of enormous energy density, totally beyond human experience. Some fundamental considerations suggested that a consistent characterization in terms of existing physical theory could be formulated after about $\Delta t = 10^{-10}$s for which the volume of interaction by then may have had a radius of perhaps 10^{10} m at a temperature of some $10^{15}\,^{\circ}$C; this interpretation implies that space and time have no meaning prior to this ignition singularity at $t = 0$ and that all matter-energy of the present universe was initially inside an incredibly minuscule volume, having an inside but no outside with a subsequent expansion of the space-time grid.

Contemporary nuclear physics suggests that during this early stage of expansion, the four fundamental forces of gravity, electromagnetism, nuclear-weak, and nuclear-strong, had separated. A profusion of particles and their anti-particle counterparts occurred when energy was converted into matter during a period of inflation; annihilation, decay, and particle-particle collisions provided further radiation energy to sustain a super-high temperature expansion. A slight excess of particles over anti-particles emerged of which the electrons and quarks proved to be of fundamental importance; in particular, the quarks combined as different triplets to form the fundamental neutron and proton, commonly represented by n and p respectively.

Though the proton (p) and deuteron $(1n + 1p)$ were the constituents of the very early universe, they were soon joined by helium $(2p + 2n)$, formed by building-block construction. Additional side reactions also occurred, producing small quantities of lithium $(3p + xn)$ and beryllium $(4p + xn)$ isotopes, each containing a specific number of protons but the number of neutrons varying between 2 and 8 to form specific isotopes of a given element. Figure 9.2 suggests these nuclear build-up processes indicating also some approximate nuclear-atomic dimensions.

Evidently then, the Big Bang triggered the build-up of increasingly larger nuclides as suggested by the heterogeneous progression.

$$energy \rightarrow \begin{pmatrix} quarks, \\ electrons \end{pmatrix} \rightarrow nucleons \rightarrow atoms. \qquad (9.1)$$

The expansion of this primeval fireball was not uniform and over the next billion years, clouds of ions formed, accretion began, and — because of radiation emission — cooling occurred. Gravity tended to condense these clouds into webs and clusters, and collective motion provided for a range of shapes, sizes, and densities. Within these clouds of ions and

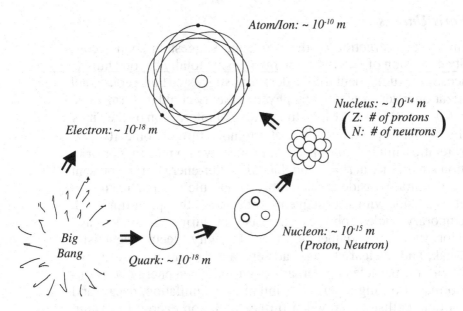

Fig. 9.2 Approximate diameters and sequential build-up pattern associated with nuclear and atomic matter.

photons, various types of collisions and decay reactions continued to occur which, when combined with non-uniform condensation, eventually formed in the order of 10^{12} galaxies each composed of an average of perhaps 10^{12} number of stars. In excess of 99% of all matter in the universe at these early times, and to a large extent even now, consist of a plasma state — that is the negatively charged electrons are separate from the positively charged nuclei and all were immersed in a low-temperature electromagnetic field.

9.2.3 *Nucleosynthesis*

Within perhaps a billion years, some stars had burned up most of their hydrogen and began burning helium thereby increasing the quantity of heavier reaction products. The cores of such stars gradually collapsed as the critical balance between outward radiation-driven pressure and inward gravity-driven pressure changed. In the most massive stars, the associated increase in the central temperature could provide conditions for an explosion and subsequent interaction with the collapsing outer proto-plasma shell, providing thereby collisional conditions for the synthesis of the heavier nuclides, up to the heaviest naturally occurring element uranium (92 protons and from 142 to 146 neutrons). This cosmic furnace is called a supernova explosion with a highly publicized event observed in the visible universe in 1987.

All matter in the universe began with a variety of reaction-driven building-block processes eventually yielding a collection of 92 elements characterized by three distinct groupings, Fig. 9.3. In addition, many of these 92 elements have both stable and naturally radioactive isotopes establishing a natural nuclear world in excess of 300 distinct nuclear species of which about 20 are naturally radioactive. All naturally existing materials are based on various compositions of these basic nuclear building blocks[†].

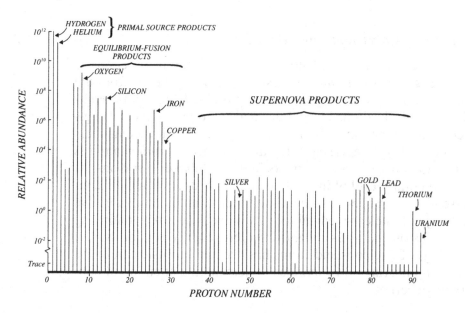

Fig. 9.3 Estimated abundance of elements in the universe indicating also stages associated with the formation of groups of elements.

9.2.4 *The Sun*

Our Sun, situated off-centre in the Milky Way Spiral Disk Galaxy, formed some 6×10^9 years ago, is a typical young star located $\sim 150 \times 10^6$ km from earth. Its central temperature is about $2 \times 10^7\,°C$ with a surface temperature of $\sim 5530°C$. It consists mostly of hydrogen ions, small

---

[†]Nuclear devices such as accelerators and fission reactors can be used to produced nuclear species not found in nature.

quantities of helium ions, and separated electrons — all held together by gravity in opposition to the thermal-plasma and radiation pressure. The fusion reactions combining light ions generate $\sim 4 \times 10^{26}$ W of power and, correspondingly, the Sun gets lighter at the rate of $\sim 4 \times 10^9$ kg/s. With a radius of $\sim 7 \times 10^8$ m and a mass of 2×10^{30} kg, its average mass density is ~ 1.4 g/cm^3 and average power density of 3 W/m^3 (i.e. like lukewarm bath water). It will evidently burn for another $\sim 10^9$ years before some 10% of its fuel is exhausted. The solar constant, defined as the average areal power density at the earth's distance attributable to the Sun is ~ 1370 W/m^2.

9.2.5   *Nucleons, Atoms, and Engineering*

The world of nucleons, ions, and atoms has had a defining effect on the evolution of engineering. It was the visible part of radiation from the atomic interaction in stellar objects which prompted Prehistoric engineers to build carefully aligned megaliths; it also prompted Ancient engineers to construct impressive observatories and led Medieval engineers to design and construct precise earth-to-star reference instrumentation as a critical tool to aid pioneering mariners in their global explorations. At a more contemporary level of interest, an understanding of nuclear concepts is essential to the operation of atomic clocks, functioning of fission reactors, design of nuclear-medical therapy devices, and the continuing challenges of fusion energy as a viable long term energy possibility. The subtle workings of nucleons and atoms has been an intrinsic part of the evolution of engineering and continues to demand the attention of contemporary engineering.

9.3   Elements and Aggregates: Terrestrial Basis

While the Big Bang model and subsequent gaseous cloud condensation — combined with stellar explosions and the capture of galactic fragments for galaxy and star formation — is widely accepted in broad terms, the emergence of individual objects such as planets in our solar system is more uncertain. The reason for this is the problem of formulating coherent cause-and-effect physical processes which account for the compositional variations of our earth and the observable features of the planets, while also explaining the existing dynamical features of our solar system. The increasingly accepted view is that accretion of stellar gas and dust is likely the most primary process.

9.3.1 *Composition*

Towards the middle of the 20th century, a comprehensive body of knowledge about wave propagation in various media emerged. In particular, an earthquake is known to generate a sufficiently intense local disturbance resulting in several types of waves which emanate outward and can be detected by sensitive surface-based pressure-wave sensing devices. Then, by methods of tomographic reconstruction based on the speed, dispersions, and refractive properties of these waves in varying media, a density distribution in the Earth can be estimated. Such seismic tomographic investigations provide convincing evidence suggestive of a dominant four-region interior of the earth:

(a) Central Core: high density metallic material (radius $\simeq 2400$ km)
(b) Outer Core: liquid metal region (thickness $\simeq 2200$ km)
(c) Mantle: less-dense material composition (thickness $\simeq 2000$ km)
(d) Crust: layer which surrounds the earth (thickness < 100 km)

Further, with the availability of Earth surface maps, it was noted that some land masses possessed mutually matching edges suggesting a past land distribution different from that of the present. By the ~ 1960s, this feature contributed to the plate-tectonics characterization of the Earth's crust with continental plates moving intermittently relative to each other at a rate of averaging $\gtrsim$ cm/year.

9.3.2 *Nuclear Effects*

Early in the 1900s the atom was recognized as a basic building block of all matter. While some atoms are stable, others undergo spontaneous nuclear decay transformation involving a mass-decrement with an attendant release of energy. This recognition provided for two geophysical consequences of significance. One is that radioactive decay in the Earth's interior provides for heat energy which diffuses to the topmost part of the mantle on which the crust of the earth is located; the resultant differential thermal forces contribute to various crustal plate-to-plate interactions, with the consequence of phenomena such as volcanic eruptions and ocean spreading. The other geophysical consequence is that nuclear decay in a collection of some radioactive *parents* would lead to specific *daughter* products, relative to a non-radioactive *companion*. Hence, knowing the rates of nuclear decay and isotopic populations at a later time could be used to determine the earlier time when a substance crystallized thereby trapping the *daughter* nuclides adjacent to its related non-decaying *companion*; experimental measurements of *daughter-companion* atom density ratios — knowing that at the time of crystallization $N_d/N_c = 0$ — gives an estimated time of $\sim 4.5 \times 10^9$ years since the formation of the Earth.

9.3.3 Geo-Processes

Considering known phenomena of matter diffusion, heat transfer, convection, and gravitational effects, one may summarize the dominant processes for the primal stages of formation of the Earth by the following:

(a) *Accretion*
    Gradual accretion of stellar matter leading to an aggregate of elements existing in the universe.
(b) *Impact*
    Continuing surface impact by small objects thereby sustaining sufficiently high temperatures to melt the metallic elements.
(c) *Compaction*
    By force of gravity and with gravitational energy converted into heat, the molten metals accumulated in the center, with the less dense gaseous elements diffusing outward to form a primitive atmosphere.
(d) *Differentiation*
    Material differentiation eventually provided for a high-pressure solid metallic inner core with a liquid metallic outer core; a thick, less dense but highly restless and very striated mantle thereupon formed, providing for a thin cool outer crust surrounding the earth.

This sequence of events for the formation of our Earth may be compactly written as a heterogeneous progression

$$\begin{array}{ccccc} \textit{accretion and} & & \textit{compaction} & & \textit{material} \\ \textit{aggregation} & \rightarrow & \textit{by gravity} & \rightarrow & \textit{differentiation} \end{array}. \qquad (9.2)$$

Figure 9.4 provides a graphical depiction of stages of the Earth's evolution while Table 9.1 lists selected data.

Volcanic activity involves the transport of molten matter from the Earth's interior through cracks in the mantle and channels between tectonic plates, while an earthquake causes sudden dislocation between tectonic plates. Mountains form when major continental plates collide. Erosion, glacier grinding, sedimentary deposition, compaction, and tidal effects also provide for other visible changes on the Earth's surface occurring typically on time-scales of centuries and longer.

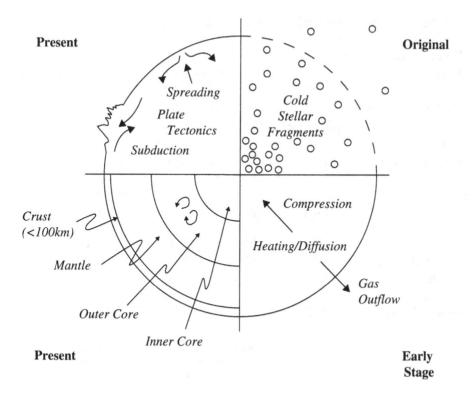

Fig. 9.4   Quadrant graph depiction of selected processes and features of the Earth: original, early and present.

It is evident that the conceptualization of the processes leading to the present form of the earth provides considerable explanatory power. For example, the outer core liquid metallic region appears to sustain considerable hydrodynamic turbulent vortex motion with the resultant dynamo effects apparently accounting for the Earth's magnetic field; indeed, the sporadic magnetic field reversal suggests possible major changes in molten metal fluidic hydrodynamic transitions while minor flow changes may account for the movement of the magnetic poles and its varying local magnetic fields intensity.

9.3.4   *Elements, Aggregates, and Engineering*

The workings of elements and aggregates provides for one of the earliest hands-on experiences for engineers: stone tools, adobe bricks, woven mats, and copper ornaments all displaying varying tactility, pliability, hardness, and tensile/compressive strength. With time, classifications

Table 9.1   Selected data characterizing the earth.

Radius (equatorial):	~6378 km
Mass:	~6.0×10^{24} kg
Avg. mass density:	~5.5 g/cm^3
Avg. earth-to-atmosphere heat flux:	~0.32 W/m^2
Avg. orbital speed:	~30.0 km/s
Avg. distance from sun:	~150×10^6 km

	R(km)	T(K)	P(MPa)	ρ (g/cm^3)
Surface:	~6378	~300	~0.1	~2.0
Crust-Mantel Interface:	~6340	~310	~0.5	~2.5
Mantel and Outer Core Interface:	~4600	~4000	~1.5	~5.5
Outer Core and Inner Core Interface:	~2400	~4600	~3.0	~10.0
Center:	0	~5000	~4.0	~13.0

were introduced and material requirements for chariots, pyramids, and water wheels could be anticipated. Ultimately, testing procedures and numerical characterizations were developed and a range of material performances under various environmental conditions could be identified. Not only do these developments relate to wear-and-tear of simple devices, but they also relate to stress-corrosion effects, resistance to extreme tectonic realignments, structures to resist natural hazards, and longevity of heterogeneous electronic functions. The workings of elements and aggregates have long been part of a subtle and persistent basis of the intellectual background of all engineers.

9.4   Monomers and Polymers: Molecular Basis

The Big Bang is the primordial trigger for the production of all ensuing forms of matter in the universe. This sequence allows an extension of (9.2) as suggested by the progression

$$energy \rightarrow quarks \rightarrow nucleons \rightarrow nuclides \rightarrow atoms \rightarrow molecules \rightarrow \cdots .$$

$$(9.3)$$

Here the notion of a building block construction process in which small components combine to yield larger and also more complex entities with a corresponding change in aggregate properties, is evident. These primal processes also lead to an increase in intrinsic organization of matter with time. What is less obvious is that not only does material reconfiguration occur but simultaneously matter-energy transformation also takes place according to

$$\begin{pmatrix} reactants\ a + b \\ possessing\ kinetic \\ energy\ E_a\ and\ E_b \end{pmatrix} \rightarrow \begin{pmatrix} matter\text{-}energy \\ collisional \\ transformation \end{pmatrix} \rightarrow \begin{pmatrix} reaction\ products \\ d + e\ having \\ kinetic\ energy \\ E_d\ and\ E_e \end{pmatrix}.$$

$$(9.4a)$$

The change in kinetic energy relates to the change in rest mass given by

$$(E_d + E_e) - (E_a + E_b) = [(m_a + m_b) - (m_d + m_e + \cdots)]c^2$$
$$= -[\Delta m]c^2. \qquad (9.4b)$$

Equation (9.4b) is the more insightful relationship of what is more popularly known as the Einstein mass-energy equivalence $E = mc^2$; in a manner of speaking, *mass is confined energy* and *energy is liberated mass*.

As suggested in the preceding sections, the material composition and processes of normal stars may be characterized by hydrogen nuclei (p), helium nuclei (He^4), and the separated electrons to provide overall charge neutrality. The helium nuclei are produced by a complex multistep reaction chain — known as the Carbon Cycle — which has the effect of yielding

$$4p \rightarrow {}^4He + energy + elementary\ particles, \qquad (9.5)$$

with the energy so released contributing to the sun's radiation emission; hence solar energy is nuclear energy. Comets are commonly depicted as dirty snowballs consisting of volatile light elements and molecules. In contrast, meteors possess a range of elemental populations of the heavier metals and low-mass molecules. Finally, our Earth possesses the full range of the 92 naturally occurring elements, Fig. 9.3, and innumerable complex molecules. There exists however a substantial difference in the elemental abundances in the earth's crust, the atmosphere, and the solar system, Table 9.2. As a point of contrast, the interstellar space possesses an exceedingly low particle density ($\sim 1\ \text{m}^{-3}$) with the background photon population at a temperature of 2.9 K ($-270°C$).

Table 9.2   Abundances of selected elements (%). Note that oxygen and silicon are the most plentiful elements on earth and that nitrogen and oxygen are the dominant constituents in the atmosphere.

Element	Z	Earth's Crust	Present Atmosphere	Solar System
H (Hydrogen)	1	0.2	0.01	71
He (Helium)	2	–	–	27
C (Carbon)	6	0.2	0.01	<0.01
N (Nitrogen)	7	0.02	78	<0.01
O (Oxygen)	8	45	21	<0.01
Al (Aluminum)	13	0.8	–	–
Si (Silicon)	14	23	–	–
P (Phosphorus)	15	0.2	–	–
K (Potassium)	19	2.9	–	–
Fe (Iron)	26	5.8	–	–
Cu (Copper)	29	6×10^{-3}	–	–
Ag (Silver)	47	1×10^{-5}	–	–
Sn (Tin)	50	1×10^{-4}	–	–
Au (Gold)	79	2×10^{-7}	–	–
Th (Thorium)	90	1×10^{-3}	–	–
U (Uranium)	92	3×10^{-4}	–	–

9.4.1 *Combinatorics*

Of wide interest is the evolution of the known 92 distinct elements into conceivable compounds by collisions. Three considerations are important:

(a) *Interactions*
    Which elements have a particular affinity for interacting? A common measure is that of a joint cross section for a particular process.
(b) *Reaction Rates*
    Which reactions are likely to have a relatively greater impact on element population changes? The concentration of elements (and substances) N_i and N_j are important because the $i + j$ two-body collisions occur at the rate of

$$r_{i+j} = \kappa_{ij}(T_{ij})N_i N_j, \qquad (9.6)$$

where $\kappa_{ij}(T_{ij})$ is a reactant temperature dependent rate parameter.

(c) *Combinations*

How many distinct reaction products are conceivable given a primal population of 92 elements? Elementary considerations suggest a maximum of $(92)^2$ possibilities. However, experiments have demonstrated that $\kappa_{ij} = 0$ for many pairings and theoretical considerations impose other restrictions — a situation becoming more restrictive with larger molecular structures.

9.4.2 *Build-up*

As elementary examples and considering a likely order of terrestrial substance formation, the following identifies reactions in which oxygen dominates:

$$O + H \rightarrow OH \quad (hydroxyl) \tag{9.7a}$$
$$H + OH \rightarrow H_2O \quad (water) \tag{9.7b}$$
$$O + O_2 \rightarrow O_3 \quad (ozone). \tag{9.7c}$$

Participation of nitrogen, sulphur, and phosphorus in reactions is suggested by

$$N_2 + 3H_2 \rightarrow 2NH_3 \quad (ammonia) \tag{9.8a}$$
$$S + H_2 \rightarrow SH_2 \quad (hydrogen\ sulphide) \tag{9.8b}$$
$$2P + 3H_2 \rightarrow 2PH_3 \quad (phosphine). \tag{9.8c}$$

Incorporation of carbon produces methane

$$C + 2H_2 \rightarrow CH_4. \tag{9.9}$$

Larger and ultimately more complex compound molecules may also appear:

$$NH_4OH \quad (ammonium\ hydroxide) \tag{9.10a}$$
$$MgSiO_3 \quad (magnesium\ silicate) \tag{9.10b}$$
$$MgAl_2Si_2O_8 \quad (magnesium\ aluminosilicate) \tag{9.10c}$$
$$CH_2(NH_2)CO_2 \cdot R \quad (amino\ acid\ R\text{-}groups). \tag{9.10d}$$

All of these chemical constituents have proven to be most vital to human life on Earth. For example, the ozone molecule O_3 in Eq. (9.7c) absorbs much of the ultraviolet part of the solar radiation which is known to be hazardous to the human skin; the small concentration of water vapor in the atmosphere scatters and reflects radiant energy contributing to the maintenance of the existing temperature range of the human habitat.

Equilibrium of any of these mono or multi-component species in the atmosphere is given by

$$\frac{dN_i}{dt} = \sum_k r_{+i,k} - \sum_\ell r_{-i,\ell}$$
$$= 0, \tag{9.11}$$

where $r_{+i,k}$ is the k-type production rate of species N_i in a unit volume and $r_{-i,\ell}$ is the ℓ-type destruction rate of N_i.

Finally and most importantly, the amino acid R-group, (9.10d), has proven to be most significant: of the some 100 different R-groups, about 40 of these are constituents of proteins which are found in all plant and animal life and are essential to human life.

Currently some 10^7 molecular compounds have been identified, many of which exist naturally and others may be produced in special-purpose chemical reactors.

9.4.3 *Catalysis*

Many important reactions are sustained by species which contribute to reactions but are themselves not transformed in the process. These are called catalysts and are widely used in the enhancement of specific inorganic reactions; biological catalysts are commonly called enzymes. A simple and general catalytic or enzymatic process is suggested by the following. Consider some chemical reactant A interacting with a catalyst C with the important reactions then proceeding by the following reaction chain:

$$C + A \xrightarrow{\kappa_{CH}} CA$$
$$CA \xrightarrow{\lambda_{CH}} CP + X \tag{9.12}$$
$$CP \xrightarrow{\lambda_{CP}} C + P$$

Here κ_i and λ_j are the appropriate reaction rate and decay rate parameters, CA is the primary $C + A$ product, CP is the intermediate CA-decay product, and P is the desired product of interest; X is a side product species. The net effect of (9.12) is evidently

$$A \rightarrow P + X, \tag{9.13a}$$

also suggested in graphical form

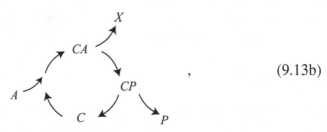

$$\tag{9.13b}$$

with each specific A initiating a cycle; thus, as long as reactants A exist and no other interruptions occur, products P and X will result. Appendix D provides for a graphical complementary interpretation of engineering-related cyclical progressions in time.

9.4.4 *Monomers, Polymers, and Engineering*

The prehistoric discovery that fire might induce smelting (i.e. decrystallization) and fusion (i.e. species production/destruction) of naturally available materials opened a Pandora's Box of possibilities. With the recognized availability of various elemental species, the elementary processes of heating and mixing soon led to results warranting even the labelling of eras — Bronze Age, Iron Age, Steel Age, and variously, the Atomic or Synthetic Age. The available choices of natural input materials combined with variations of pressure, temperature, sequencing, electric/magnetic fields, and processing time soon led to a myriad of monomers and polymers all possessing distinct properties for different purposes; interestingly, one has yet to find the limit of such outcome diversity. No engineering discipline is unaffected by the consequence of the range of monomeric and polymeric materials.

9.5 Cells and Biota: Biological Basis

Palaeontologists of this century have identified single-cell microfossils and aqueous fauna as suggestive of the earliest and most elementary form of life appearing on Earth perhaps 10^9 years ago, and these were then followed by hard shells and waterborne vertebrates appearing about 10^8 BP. Subsequently, about 10^7 BP, a profusion of flora and fauna appeared on land, in water, and in the atmosphere providing an estimated $\sim 10^7$ distinct types of living species. There exist indications that extreme volcanic activity combined with asteroid impacts may have occurred in the past destroying entire families of animal and plant life; the most severe such extinctions are estimated to have occurred about 250 and 65 million years ago with the former causing the extinction of some 90% of species on Earth and the latter causing the demise of the dinosaurs.

9.5.1   *Bio-Populations*

Important to the early understanding of fundamental aspects of biology — that is the processes associated with living things — were the observations of microscopic sacs of fluid in human tissue, first observed with crude microscopes more than 100 years ago. These soft, lumpy objects now called cells, were about 10^{-5} m in diameter and possessed a most remarkable feature: they multiplied by pair division. This notion of a population increasing by self-division was soon recognized as totally different from non-biological population increases based on collisions between distinct entities such as nucleons, atoms, or molecules. These distinctions are illustrated in the following for the increase of some I-type population of entities in successive generations displaying also the dynamics of inorganic and human-made manufactured populations for comparison:

(a) Biological population increases by self-division:

$$i \rightarrow 2i \rightarrow 4i \rightarrow 8i \rightarrow 16i \rightarrow \cdots \qquad (9.14a)$$

(b) Inorganic population increase by interaction:

$$l + m \rightarrow i + \cdots \qquad (9.14b)$$

(c) Manufactured population increase by addition:

$$i \rightarrow 2i \rightarrow 3i \rightarrow 4i \rightarrow \cdots . \qquad (9.14c)$$

Their corresponding growth or decline rates now possess an important subscript and component distinction. Using $r_{\pm i}$ as an expression for the rate of growth or decline of a population N_i, the rate proportionality relations for the above population changes, (9.14), is suggested as follows:

(a) Biological population:

$$r_{\pm i} \propto N_i \qquad (9.15a)$$

(b) Inorganic population:

$$r_{\pm i} \propto N_l N_m \qquad (9.15b)$$

(c) Manufactured population growth:

$$r_{\pm i} \propto constant. \qquad (9.15c)$$

Herein resides a profound distinguishing feature of the phenomenon of life and the cell was soon recognized as deserving to be studied in more detail.

9.5.2 *Cell Details*

It is estimated that about 10^{14} cells exist in a human adult body and that the average time between cell divisions can vary from hours to days. These cells absorb nutrients and produce a range of complex products. Indeed, to use an analogy, the workings of a cell possess features of a micro chemical reactor or a micro factory in which various inflow nutritional compounds are broken down and then reassembled to form a host of polymers including proteins. The proteins produced consist of very long molecular chains containing amino acids which were soon recognized as most important. Of the several hundred types of proteins in the body, many are critical for purposes such as the suppression of specific infections, hormonal production, healing of wounds, metabolism, and other specific human biology functions.

A more detailed examination of the cell has identified a central domain, called the nucleus, from which emanate thin tendrils with attached ribosomes, Fig. 9.5. While protein production occurs in these ribosomes, the nucleus appears to be the command centre which provides instructions for the type of proteins to produce, not unlike directives in a factory emanating from head office.

Beginning in the 1930s it was shown that the nucleus of the human cell contains 46 (23 pairs) chromosomes composed of various proteins and the deoxyribose nucleic acid, a complex polymer commonly referred to by its acronym DNA. The characterization of this DNA — a 2 m long and 0.01 μm thick spiral with a self-copying structure and

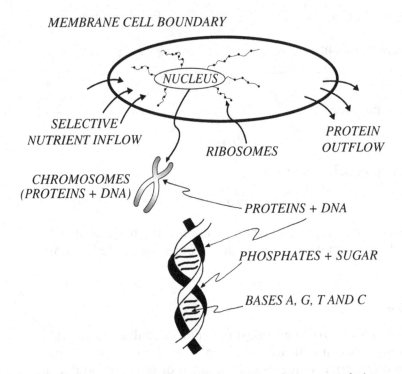

MEMBRANE CELL BOUNDARY

NUCLEUS

SELECTIVE NUTRIENT INFLOW

RIBOSOMES

PROTEIN OUTFLOW

CHROMOSOMES (PROTEINS + DNA)

PROTEINS + DNA

PHOSPHATES + SUGAR

BASES A, G, T AND C

Fig. 9.5 Schematic depiction of selected components and processes of a human cell.

function — proved to be a stunning achievement of creativity involving intuition and X-ray diffraction devices of the 1950s. Its discoverers were James Watson, USA, and Francis Crick, England; together with Maurice Wilkins, England, they shared the 1962 Nobel Prize for this work. In summary, DNA possesses a double helical structure consisting of

(a) Phosphates
(b) Sugars, and
(c) Bases of four types: adenine (A), guanine (G), thymine (T), and cytosine (C).

The phosphate and sugar molecules were found to make up a helical backbone from which the four bases project, Fig. 9.5. These bases then match-up — like rungs of a ladder — with other sets of four bases attached to a companion phosphate + sugar backbone. Two phosphate and sugar strands spiral around each other separated by matching base rungs defining the famous double helix.

Several features of DNA were soon found to be of critical importance in human biology:

(a) One DNA possesses approximately 3×10^9 base pairs and is distributed — in varying length — among the 46 chromosomes.
(b) There exist specific patterns for the matching of bases from one strand with the other.
(c) A minimum length of distinct base sequences along a strand of DNA carries the code for protein production in the sequence of bases; this DNA section is called a gene and the set of genes characterize the hereditary uniqueness of a human. It is estimated that a human contains some 10^5 such basic genetic units of varying length, though only about 3% of base pairs are involved in the gene structure.

The most notable feature of a human cell resides in its capacity to self-replicate: a DNA will break apart by splitting between two base residuals, with each half of the DNA then selecting the correct matching bases and associated phosphate and sugar backbone from its cell environment to form two complete and identical cells from its original.

Based on the above considerations and using the idea of heterogeneous progression notation, one may now also write

$$atoms \rightarrow molecules \rightarrow cells \rightarrow polymers \rightarrow \left\{ \begin{array}{c} self\text{-}organizing \\ biological\ functions \end{array} \right\}$$

$$(9.16)$$

noting that while atoms and molecules are the primary compositional entities, the end product now relates to specific functions uniquely associated with biological organisms, that is life with its distinct processes and functions:

$$\{biological\ functions\} = \left\{ \begin{array}{ll} replication & excretion \\ metabolosm & memory \\ self\text{-}repair & coordination \\ regulation & catalysis \\ modification & conversion \\ extinction & homeostasis \\ ingestion & \vdots \end{array} \right\} . \qquad (9.17)$$

These functions resulting from the selective organization of naturally occurring components represents the great watershed: the existence of living things.

9.5.3 *Infection*

The above characterizations can be used to establish important features
associated with disease:

(a) *Bacterial Infection*
    Disease may occur when the protein products are unable to perform
    their biological functions because of the presence of disruptive bac-
    terial agents, called bacterial pathogens.
(b) *Viral Infection*
    Disease may also occur when a disruptive viral agent — called viral
    pathogens — enters a cell causing the production of faulty cell reac-
    tion products.

Figure 9.6 provides a schematic of these two processes. Note the resem-
blance of these biological processes to common industrial processes: an
engineered device may fail in use because of damaging *external* interference (i.e. bacterial infection) or because of faulty *internal* manufacturing
(i.e. viral infection).

The treatment of bacterial infections — e.g. typhoid, pneumonia,
cholera, diphtheria, … — is commonly undertaken with antibiotics such
as penicillin in order to neutralize the effects of bacterial pathogens.

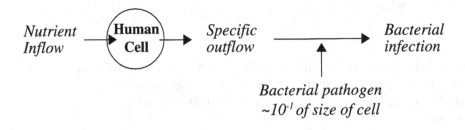

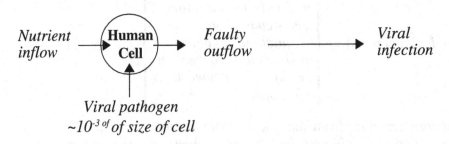

Fig. 9.6   Comparison models of infections.

Apprehension about the excessive use of antibiotes has recently emerged because of the appearance of bacterial strains which are increasingly resistant to such treatments.

Medical treatment against viral infection is only partially successful. Vaccination against smallpox, chickenpox, measles, tuberculosis, and polio has proven to be effective but treatment for other viral infections such as herpes, HIV, SARS, and even the common cold, is less effective.

Beneficial bacterial actions, also known as fermentation, occur widely; among these is the production of food (e.g. cheese, butter, yogurt, wine, ...), digestion in humans and animals, vegetation composting, nitrogen fixation in plants, and many others. Bacterial processes are indeed essential to nature's balance.

9.5.4 *Applied Biology*

The branch of biology which involves splicing, recombining, and selective rearrangement of specific parts of DNA is commonly called genetic engineering or genetic modification (GM). There now exists a commercial market for genetically altered organisms which, to many, is equivalent to conventional historical selective breeding but to others a potentially risky precedent. Since the late 1980s, most developed countries have established or licensed special laboratories for genetic alteration with its activity guided by a variety of restrictions.

Forensic science, biological research, and biotechnology have made substantial use of so-called DNA fingerprinting: that is, the determination of patterns of the bases in DNA uniquely associated with specific organisms. This involves processes and devices readily expressible in engineering terms. To begin, a small specimen in which cell formation has taken place — blood, bone, skin, semen, hair, etc. — is cleaned and chemically treated to break open the cells, with the DNA then separated in a centrifuge. Enzymes which recognize the DNA base pattern break the helix at specified points providing DNA molecules of varying lengths as determined by its specific component genes. With DNA negatively charged, their placement in a gel situated in an electric field will subsequently lead to DNA ordering according to their gene length. A nylon blotting material will absorb these DNA molecules which then combine chemically to a known synthetic and radioactive DNA for bonding to their associated base sequences. By means of contact autoradiography, the uniqueness of a DNA pattern becomes visible in a standardized format.

9.5.5 *Biota, Cells, and Engineering*

Engineering has a specialized participatory history with human biology. This involvement began about 400 years ago with the design and adaptation of non-invasive prostheses related to human mobility; here, engineers first encountered the complexities of interfacing common engineering materials and devices with the special biotic requirements of human skin, muscle, and bone. In more recent times, engineers extended this experience to the design and manufacture of implantable bone-reinforcing bolts and stimulating heart-pacers. This success led to metallic joint replacements and other inserts, demanding however that engineers now acquire an increasing understanding of the biological environment and its workings. Recent engineering work has led to sophisticated devices uniquely adapted to cell and DNA analysis requiring an engineering understanding of the workings of biota and cells.

9.6   Precursors to Engineering

Our discussion of primordial evolution of nature evidently suggests a heterogeneous progression triggered by the Big Bang as

$$
\begin{array}{cccc}
astronomical & \to & terrestrial & \to & molecular & \to & biological \\
(\sim 10^{10}\,\mathrm{BP} \to) & & (\sim 10^{9}\,\mathrm{BP} \to) & & (\sim 10^{8}\,\mathrm{BP} \to) & & (\sim 10^{7}\,\mathrm{BP} \to)
\end{array}
\tag{9.18a}
$$

or, with a different focus,

$$
\begin{pmatrix} nucleons, \\ electrons \end{pmatrix} \to \begin{pmatrix} atoms, \\ molecules \end{pmatrix} \to \begin{pmatrix} monomers, \\ polymers \end{pmatrix} \to \begin{pmatrix} cells, \\ biota \end{pmatrix}.
\tag{9.18b}
$$

Evidently this sequence embodies an enormous number of processes and distinct entities, some of which are suggested by a sequential decomposition of $N(t)$:

$$
N(t): \quad
\begin{pmatrix} Explosion \\ Inflation \\ Nucleosynthesis \\ Stars \\ Planets \\ \vdots \end{pmatrix}
\to
\begin{pmatrix} Cooling \\ Atmosphere \\ Mountains \\ Rivers \\ Oceans \\ \vdots \end{pmatrix}
\to
\begin{pmatrix} Monomers \\ Catalysis \\ Polymers \\ Oxides \\ Acids \\ \vdots \end{pmatrix}
\to
\begin{pmatrix} Cells \\ Replication \\ Self\text{-}repair \\ Adaptation \\ Populations \\ \vdots \end{pmatrix}.
\tag{9.19}
$$

This heterogeneous progression also illustrates the important feature of initiation and simultaneity of progeny dynamics. For example, while the initial Big Bang expansion (col. 1) preceded cell formation (col. 4), stellar expansion (col. 1) still continues to this day; similarly nucleosynthesis (col. 1) still takes place even though geophysical changes continue (col. 2) and catalysis (col. 3) is taking place in a range of environments. With each of the four process domains still continuing, nature is evidently still evolving.

The discussion in this Chapter has focused upon selected evolutionary aspects of astronomical (electrons, nuclei), terrestrial (elements, aggregate), molecular (monomers, polymers), and biological (cells, biota) features of nature associated with the $\sim 10^{10}$ BP to $\sim 10^{6}$BP time interval. This heterogeneous progression also illustrates the natural foundational connections to contemporary Engineering disciplines. Among these, one may cite specifically the essential linkages of electrons with Electrical Engineering, molecules with Chemical Engineering, aggregates with Civil Engineering, gases and metals with Mechanical Engineering, nuclei with Nuclear Engineering, biota with Biomedical Engineering, and others.

Additionally, it is recognized that specializations in engineering disciplinary classifications may come about by the judicious integration of selected parts of physical reality in the creation of new and ingenious devices. An acute awareness of natural entities and processes within an interconnected and dynamical context can evidently identify further new devices of critical importance to the continuing adaptive evolution of engineering.

In a physical-historical sense, one may well assert that nature $N(t)$ constitutes the precursor to all engineering theory and practice; that is

$$N(t) : \left\{ \begin{array}{c} Precursor\ to \\ Engineering\ in\ Time \end{array} \right\} \quad (9.20)$$

may be taken as a most foundational basis for engineering and graphically integrated with the seven engineering eras described in Chapters 2 to 8 and graphically suggested in Fig. 9.7.

This graphical depiction also suggests the intrinsic distinguishing systematics of three classes of phenomena:

(a) One postulated explosive singularity trigger
(b) Three S-shaped {birth} → {growth/maturation} progressions

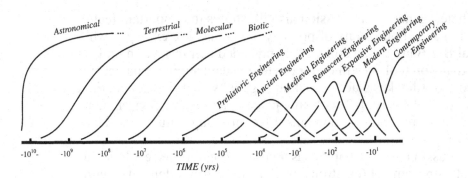

Fig. 9.7 Schematic depiction of the evolution of engineering precursors and ensuing engineering eras. The evident diversity of phenomena here displayed makes numerical coordinates on an ordinate axis impossible.

(c) Seven bell-shaped {emerge} → {saturate} → {decline/substitute} progressions now involving human creative thought and skilled actions directed to a societal interest.

Note that each of these progressions has been discussed in text.

9.7 Nature and Human Response

The preceding emphasis on selected natural entities and phenomena corresponds to a widely accepted contemporary scientific interpretation of physical reality. There is, however, more to nature than its scientific or strictly objective basis with the more commonly encountered subjective human response as especially relevant to many aspects of engineering.

With considerable generality the dominant human response to nature can be organized into two partly overlapping subjective and highly personal categories:

(a) *Transcendental engagement* with nature
(b) *Sense-of-place attachment* in nature

Transcendental engagement with nature has its origin in a sense of awe and delight about the apparent incomprehensibility and innate appeal of natural entities and associated processes. This sense of subjective wonder and amazement leads to religious, aesthetic, and philosophical streams of thought.

Religious considerations relate to the establishment of a human association with nature in a more inclusive sense; one such connection may involve a belief of Nature as a Spirit Mother which supplies all essential human survival needs while another is based on the belief of Nature as the product of Divine Creation involving humans as active though temporary custodians of the terrestrial domain. Aesthetic responses seek various forms of literary and artistic expressions as a means of inducing a sense of deep and possibly spiritual affinity with the natural world.

The philosophical response is largely intellectual; herein, one seeks to establish a more basic perspective of natural reality with functional harmony and logical consistency often as critical features. These and related considerations have established several branches of philosophy of which the following possess particular relevance to engineering:

(a) *Epistemology*
Theory of Knowledge (e.g. How do we know what we know? Are some phenomena beyond comprehension or is it simply human ignorance? ...).
(b) *Metaphysics*
Theory of First Principles (e.g. What is the original cause of observed dynamical effects? Why do scalar density gradients induce vector currents? ...).
(c) *Ethics*
Theory of Rational Behavior (e.g. Why should one act in a particular way? Are there special principles which should govern an engineer's actions under a given set of conditions? ...).

In contrast to *transcendental engagement*, the phrase *sense-of-place attachment* seeks to convey a degree of human attachment to some geographical locality and engender subjective feelings of *belonging to* and have *having some expectations about* a local area called *place*. A particular facet of this sense-of-place is already suggested by the Cro-Magnon cave paintings as visible depictions of their attachment to a place. Subsequent Sumerian ziggurats and later Egyptian pyramids similarly projected an identification of a people with some region. All cultures hold burial ground as sacred and individuals invariably judge ancestral residential domains and historical places as deserving of special recognition.

With time, the sense-of-place concept led to the establishment of physical structures such as fortification and the use of rivers as perpetual lines of demarcation between adjoining people places. Successive reduction of the size of places led to tree lines and fences signifying ownership, as well as mutually agreed upon shared places. Finally, in Modern times,

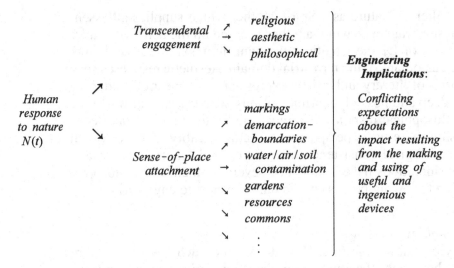

Fig. 9.8 Precursor connectivity associated with human response to nature and its engineering implications.

this idea of a sense-of-place became identified with legally empowered possession and listed ownership rights.

Violation of transcendental engagement and sense-of-place had its subtle beginning with ancient diggings of irrigation channels and dam construction: the making of some devices which enhanced some local place tended to violate some other place by altering an expected flow of water. These transboundary conflicts took on new dimensions in the industrial era when localized hydro-carbon combustion led to regional and increasing atmospheric contamination by wind action. A contemporary expression of many similar intrusive effects is currently implied by the acronym NIMBY (*Not In My Back Yard*).

Relation (9.20) summarizes a connected depiction of human response to nature and its extension to a common vexation of immediate interest to engineering, Fig. 9.8; there is no escaping this dilemma and the only reasonable route for engineers to take is to become increasingly conversant with the more basic issues arising from the heterogeneous engineering connectivity involving $N(t)$.

9.8 To Think About

- Estimation of the time of the Big Bang requires independent measurement of the recessional speed and distance of stellar objects. How do such measuring devices function and which of these two types of measurements is the more difficult?

- Describe some of the devices used to mitigate some adverse effects of nature's self-dynamics: floods, earthquakes, hurricanes, etc.
- Much of engineering research involves Direct (or Forward) problems; that is, one induces a controlled cause and then assesses the effect. Indirect (or Inverse) problems are also important particularly in the assessments of natural phenomena; in these problems the order is reversed, one observes an effect and wishes to obtain some details about the cause. Identify several such Indirect problems of relevance to engineering.

CHAPTER 10

Engineering: Patterns and Specializations

$$N(t) \to E(t) \to D(t) \to \cdots$$

10.1 Embryonic Forces

Engineering, that is the thoughtful and skillful way of using natural materials and phenomena in the making of useful and ingenious devices,

$$N(t) \begin{Bmatrix} matter \\ phenomena \end{Bmatrix} \to E(t) \begin{Bmatrix} creative\ thought \\ skilled\ actions \end{Bmatrix} \to D(t) \begin{Bmatrix} ingenious \\ useful \end{Bmatrix} \tag{10.1}$$

emerged in primitive form in the early Paleolithic era $\sim 10^6$ BP when humans first began collecting stones to be chipped for particular purposes. While this prehistoric form of surface mining and impact manufacturing involved readily available inorganic materials, human inventiveness subsequently established recursive variations on this basic activity: clubs were shaped from tree roots, sharp-edged stones were attached to tree branches with animal sinew to form chopping tools, chipped stones were fastened to the end of wooden rods to create spears, and so on. Searching, testing, shaping, and synthesizing using various materials for purpose of creating useful things was initiated by Stone Age ingenuity and has continued to be an enduring feature of engineering.

Throughout most of its history, engineering evolved by direct and empirical means. Natural materials were extracted, manipulated in various ways, combined to yield distinct geometrical configurations, and then tested for their functionality. As necessary, changes were introduced and the iterative cycle became more specialized as engineering became increasingly science-based and more professional. Two primary forces provided the continuing impetus for this enduring activity:

(a) *Creation Push*
    It is an intrinsic and enduring human trait to create. For engineers, this involves inventing new devices and altering existing devices — even before a comprehensive mechanistic understanding of the device had become established.

(b) *Acquisition Pull*
    Individuals and communities possess an innate and attracting interest in new and altered devices — even before the full societal impact of the device has been recognized.

These two deficiencies notwithstanding — that is, an incomplete understanding of the workings of a device and an insufficient understanding of its adverse side effects — a variety of incentives and community imposed regulations have been introduced over time to stimulate the processes of creating novel devices subject to bounded selection by society.

This complementarity between individual creative talents and community interests has served to provide for some overarching descriptions of engineering-in-time suggested in the following heterogeneous progression:

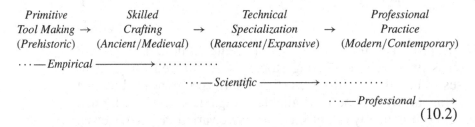

$$(10.2)$$

Evidently the empirical tradition was dominated by direct sense-experience, the scientific focus introduced a detached bottom-up theoretical and experimental methodology, and the more recent professional emphasis suggests a top-down obligatory perspective with an orientation towards a broader set of economic, political, and religious considerations.

10.2   Organizational Features

The earliest organizational features which related to those engaged in the making of ingenious and useful devices occurred in Prehistoric times when scribes, artisans, and builders were evidently set apart in recognition of their particular skills. An initial social stratification thus became established and the works of early engineering became recognized. Subsequently a master-apprentice structure emerged where one became an engineer by observing, imitating, and eventually taking the initiative in the working with and making of devices. Indeed, this apprenticeship system has survived in selected trades to the present.

10.2.1 *Engineering Education*

A significant stimulation to engineering education can be traced to the mid Renaissance by the creative conceptualizations of Leonardo da Vinci (1452–1519), by the initiative and intellectual independence of Nicolaus Copernicus (1474–1543), and by the combined theory-experiment approach of Galilei Galileo (1564–1642). The parallel developments in expanded mining, increased iron-making, extended road and canal construction, and the publication of engineering handbooks, tended to bring together theory and practice. France early recognized the implications of this integration for military purposes and introduced an advanced educational program for military engineering — the Corps of Military Engineering — about 1675 and in 1720 extended this academic emphasis to civilian road and bridge engineering. It was, however, in 1794 that the important concept of a polytechnical education for engineers — that is, a university level institute often simply called a Polytechnic — was introduced, again in France, and suggested a general model for university level engineering education; as a professional parallel, the Institution of Civil Engineers was founded in London in 1818. Formal engineering education and engineering professionalism — with considerable national variations — was now underway.

Shortly after its establishment in 1802, the United States Military Academy at West Point developed an expanded technical focus and came to be the principal route to engineering practice in the USA at the time. Then Norwich University, Norwich, CT, and Rensselaer Polytechnic Institute, Troy, NY, founded in 1821 and 1824 respectively, were first to offer formal civilian engineering programs; the first university engineering degree in the USA was awarded by Rensselaer in 1835.

The general evolving trend in organized learning eventually led to three dominant sponsorship and regulatory patterns for educating engineers:

(a) *State/Provincial/Regional Ministries of Education*
    The educational program for engineers is part of a broadly based government supported program of an autonomous degree granting

university in which a Faculty or College of Engineering is one of several administrative components; North America, Western Europe, and many other countries have adopted this educational model.

(b) *National Special-Interest Ministries/Departments*

In this model, particular ministries of a federal government are assigned special responsibility for educational institutions of relevance to their interest. For example, a Ministry of Transportation may sustain separate or combined university level institutes of Highway and Railway Engineering, while a Department of Energy may provide direct support and control over a College or Department of Hydroelectric Power Engineering. Such a structure is generally common in Eastern Europe and Asia.

(c) *Commercial and Private*

Alternatives to the above classifications exist such as commercial universities and privately endowed universities.

10.2.2 *Engineering Profession*

An important feature of modern times has been the emergence of *professions*. This term commonly refers to a vocation characterized by the following:

(a) It is based on a coherent body of theoretical and experiential knowledge.

(b) It possesses an organizational structure which subscribes to a legally empowered regulatory regime concerning member admission and discipline.

(c) Its members exhibit an identifiable domain of specialized competence and seek to provide a service function in the public interest.

A rationale for the establishment and sustainment of a profession is based on social status and on an inspired work ethic emanating from a coordinated and collective promotion of a specialized service in the public interest. Medicine, Law, and Engineering are among the more widely known professions.

Engineering professions are generally self-regulating organizations governed by National or Regional statutes which spell out details on Admission Criteria, Standards of Good Practice, and Disciplinary Procedures. These are augmented by Professional Guidelines involving Codes of Ethics, Limitations on Practice, and Designations of Labels. National and international umbrella organizations are often established to provide guidance on educational requirements as well as on transfer-ability of Licence to Practice.

A licensed or registered engineer designation can generally be acquired by a person with at least four years of university level education in a program of study recognized by an appropriate accreditation body; additionally, special examinations related to the field of practice and typically four years of satisfactory work experience under the supervision of a professional engineer, are commonly required as the final stages. In North America, the suffix-title designation such as Professional Engineer (P.Eng. or P.E.), Licensed Engineering (Lic. Eng.), or Registered Engineer (Reg. Eng.) is legally recognized while in Europe the label Diplom Ingenieur (Dipl. Ing.) or Ing. — the latter as a legitimate prefix to a name — has been adopted. Note that it is important to distinguish between an engineering degree bestowed by a university and a professional engineering designation granted by a licensing body, though the former is generally a prerequisite for the latter.

While some employers may expect their engineers to be professionally registered, this is often left to the individual. However, self-employed engineers, consulting engineers, and employed engineers whose work often places them in direct public contact, generally find professional registration advantageous.

10.2.3 *Engineering Classifications*

The classification of engineering theory and practice is universally recognized as important. For example, it serves to identify basic administrative units for universities (e.g. Civil Engineering, Mechanical Engineering, etc.) and is critical to the establishment of qualifying criteria for professional licensing. It is also essential for purpose of strategic workforce planning by industry and governments, serves well to characterize technical societies in support of professionalism, and aids in the maintenance of specialized journals, thereby promoting individual competence and inventiveness.

While a classification of the domain of engineering as academic programs and professional practice are generally recognized as vital, the establishment and sustainment of such categories together with the delineation of the associated boundaries of distinction, has long been problematic. The obvious reason for this rests in the intrinsic birthing dynamic of technological evolutions: engineering classifications arise simultaneously with changing perceptions about aspects of nature and the emergence of some technical opportunity of promising importance to society. Inevitably, labels of designation needed to be assigned and boundaries established even though the subject's continuing evolution may eventually transcend the initial linguistic designations. The common criticism that some engineering classifications and labels eventually became outdated has a credible basis.

A particular North American engineering discipline distribution pattern is readily established where an evolving system of accrediting university programs of engineering has long been in existence. A graphical depiction of accredited programs and the frequency of their occurrence is suggested in Fig. 10.1. As is evident, and within some implied uncertainty bounds attributable to regional variations in program names and content, the classification of Fig. 10.1 suggests three cluster groupings:

(a) Chemical, Civil, Electrical, and Mechanical
(b) Computer, Industrial, Metallurgy, etc.
(c) Naval, Systems, etc.

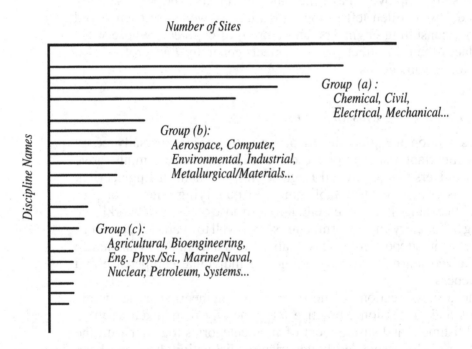

Fig. 10.1 Abundance of accredited engineering programs in North America in the mid 1990s. Listings in each group are in alphabetical order with ~350 as the maximum number of sites. However, an uncertainty of about ±10% in the relative number here displayed is implied because of variations in classifications and apparent multiple or combined programs.

Distinctions between these three clusters of the engineering profession becomes more pronounced if student enrolment is considered. Typically, the four disciplines of Group (a) enroll about 65% of all engineering students.

With these classifications, some obvious details can be included in the engineering connectivity

$$N(t) \rightarrow E(t) \begin{Bmatrix} Chemical \\ Civil \\ Computer \\ Electrical \\ Mechanical \\ Naval \\ Nuclear \\ \vdots \end{Bmatrix} \rightarrow D(t) \rightarrow S(t) \rightarrow R(t) \quad (10.3)$$

again suggesting a vector property of $E(t)$.

10.2.4 *Retrospective on Classification*

The history of engineering classification is often described in terms of specific technical developments and events. There exists, however, an underlying repeating pattern. This pattern appears to be characterized by the spontaneous emergence and synthesis of two foundational notions, one discovered and the other invented, Table 10.1:

Table 10.1 Listing of discovered and invented device developments which led to the establishment of four engineering disciplines.

Engineering Discipline (Time of Emergence)	Phenomenological Basis (Discovered)	Engineered Devices (Invented)
Civil (~3000 BCE)	Statics (stationary matter)	Structural Devices
Mechanical (mid 1700s)	Dynamics (motion of objects)	Machined Devices
Chemical (early 1800s)	Matter Transformation (molecular interactions)	Processing Devices
Electrical (mid 1800s)	Charge Transport (electron mobility)	Electro-Magnetic Devices

This tabular characterization evidently suggests a qualitative technical niche description for specific engineering disciplines which have become firmly established. Further, as a basis for each of these niches, there also exists a useful quantitative measure which provides an objective physical criteria to distinguish the above four disciplines from each other.

It is generally recognized that engineering has progressed in concert with a reductionistic and microscopic developmental perspective of physical reality. That is, a sequential historical process of increasing understanding of nature and its forces involved increasingly smaller scales of objective measures. This was evidently accompanied by tools allowing for smaller measurements of phenomena and entities in terms of the basic units of length, time, mass, electric current, temperature and luminosity. These tools and associated insights were consequently utilized for practical purposes by the engineers of the day.

A particular measure may be associated with each of the engineering developments of Table 10.1 Specifically, one may consider the range of a *minimum-length* which proved to be critical in relating the discovered phenomena to defining a family of fertile devices associated with the four engineering disciplines of Table 10.1. This numerical niche measure is displayed in Fig. 10.2 and listed in Table 10.2.

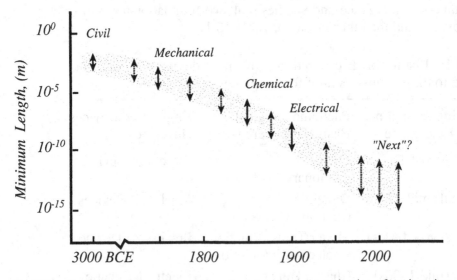

Fig. 10.2   Depiction of the *minimum-length* scale as a characterization of engineering disciplines.

Table 10.2   Listing of order-of-magnitude of minimum-length measures associated with four engineering disciplines.

Primary Engineering Discipline	Defining Characteristic Physical Phenomena	Approximate Range of Minimum-Length (m)
Civil (~3000 BCE)	structural component dimensions	10^{-3} to 10^{-2}
Mechanical (mid 1700s)	mean-free-path of molecular motion	10^{-5} to 10^{-3}
Chemical (early 1800s)	molecular interaction distances	10^{-8} to 10^{-5}
Electrical (mid 1800s)	atomic conglomerate dimensions	10^{-9} to 10^{-7}

The feature of creative thought and skilled actions of engineers are evidently comprehensively promoted and developed by their organizational features. Hence, we may introduce an engineering connectivity detail by writing

$$N(t) \rightarrow E(t) \left\{ \begin{pmatrix} thought \\ actions \end{pmatrix} : \begin{pmatrix} education \\ classification \\ professionalism \end{pmatrix} \right\} \rightarrow D(t) \rightarrow \cdots$$

(10.4)

as a point of symbolic description.

10.3   Diversity in Engineering

Tables 10.1 and 10.2 and Fig. 10.2, display aspects of the historical evolution of engineering disciplinary classifications listing only the dominant four disciplines. Various additional engineering disciplines have emerged over time and need to be noted for their importance to the continuing evolution of engineering.

10.3.1   *Niche Identification*

The graphical depiction of the dominant four disciplinary groups in Fig. 10.2 suggests a number of gaps which could well be filled by other engineering specialization. For example, Naval Engineering and Geophysical Engineering may be identified in this category of engineering diversification by specialization.

10.3.2 *Continuation by Reduction*

It is evident that the process of scientific reductionism has not ceased with the appearance of the last member of the four dominant disciplines (i.e. Electrical Engineering, mid 1800s) and one may well examine scientific and technological developments which have subsequently occurred in order to cautiously explore the conceivable appearance of a *Next* member in Fig. 10.2.

The earlier part of the 20th century has been characterized by a number of specific scientific developments. Those which most clearly followed the path of reductionism in the physical sciences and which have influenced social life, and technological developments are associated with the increasingly utilitarian fine structure of the atom and its nucleus. Several areas in which these developments have already led to a range of new devices of social utility are lasers, fission energy, fiber optic devices, medical diagnostic/therapeutic devices, new materials produced by ion beam technologies, industrial nondestructive testing facilities, nano-structures, plasma torches, and ion engines. Other areas such as fusion energy and atomic-scale construction and processing offer interesting continuing developments at this level of atomic-nuclear fine structure.

A unique and distinguishing feature of this emerging *Next* domain of engineering theory and practice is the essential need for the frequent use of a particle-wave duality and quantum-transition description of the underlying physical phenomena and corresponding nano-scale dimensions of associated device components.

10.3.3 *Specialization by Division*

The preceding lists an engineering disciplinary evolution based on the notion of niche identification and phenomenological reductionism. Additionally, a trend of the late 20th century has been an increasing specialization by division of existing engineering disciplines and variously labelled as departmental options or departmental specializations. Requiring that these specializations must also meet the same minimum-length criterion of their parental discipline suggests the following disciplinary specializations:

(a) Civil: {Earthquake Eng., Water Resource Eng., ...}
(b) Mechanical: {Aerospace Eng., Manufacturing/Industrial Eng., ...}
(c) Chemical: {Petroleum Eng., Polymer Technology, ...}
(d) Electrical: {Communication Eng., Systems Control, ...}

10.3.4 *Integration by Selection*

Recent trends have also led to the identification of interdisciplinary disciplines. In the classifications already listed, one may identify those areas of engineering theory and practice which — in order to more fully serve a particular societal interest — relate to a technical subject on several minimum-length scales, Table 10.2. Some evident integrations are the following:

(a) Agricultural Engineering: aspects of Civil, Chemical, and Mechanical Engineering.
(b) Biomedical Engineering: aspects of Chemical, Electrical, Mechanical and *Next* Engineering.
(c) Environmental Engineering: aspects of Civil, Chemical, Mechanical and *Next* Engineering.

Additional interdisciplinary programs may be identified. Also, thematic engineering programs such as Engineering Science, Engineering Mathematics, and Systems Engineering, could similarly be recast as interdisciplinary programs.

10.3.5 *Detailed Wrinkles*

Pluralistic social and industrial influences which affect engineering education and practice may — at various times — evolve into ambiguous and inconsistent situations. A present and important case-in-point is the designation of Computer Engineering with two implicit meanings: one is that of *Computer* as a hardware artifact for design, testing, and manufacture — in which case it is better categorized as a specialization in Electrical Engineering — and the other meaning is that of *Computer* as a software tool (e.g. coding, data storage, computation, communication, word processing, information management, programming, algorithm access, etc.) in which case current professional requirements demand it to be part of every engineering students educational experience; that is, all engineering programs require a *computer* as an instructional subject and working tool of relevance to the chosen disciplinary program — without a need for explicit labelling. A third categorization holds that *computer* or *software engineering* is deserving of a separate disciplinary characterization.

Overlapping areas of Mining-Mineral-Metallurgical-Materials Engineering may also be clarified by reference to the minimum-length criterion of Table 10.2. Evidently, the structural parts of Mining would seem to belong to Civil Engineering and the extractive and processing parts of the Mineral-Metallurgical subject areas are consistent with aspects of Chemical Engineering. Some parts of Materials Engineering may well be placed into the Electrical Engineering classification while others, those of relevance to optical and nuclear technical interests, would seem to be better housed in the *Next* engineering discipline. In general, some functional and dimensional aspects of materials engineering would seem to belong as a fundamental subject into every engineering discipline.

Finally, one may note extended programs. This classification refers to those academic programs best served by the addition or superposition of an equivalent of a full year of study of some allied field of professional reference. An example is Engineering Discipline X + Management Science or Engineering Discipline X + Social/Cultural Studies.

10.4   Natural Resources

The subject of this chapter, suggested also by the subtitle $N(t) \rightarrow E(t) \rightarrow D(t) \rightarrow \cdots$, evidently relates to the thoughts of engineers as they seek to connect nature $N(t)$ to specifics of devices $D(t)$. This connection may involve a range of accessible natural materials for device purposes of interest. Though details of the extraction of terrestrial materials has changed greatly from the earliest practices of surface mining for stone tool making, contemporary engineers are now confronted with exceptional questions: the conceivable local depletion of modest-cost non-renewable resources and the delicate dynamics of renewable resources.

We first consider selected aspects of this $N(t) \rightarrow E(t)$ resource renewable/non-renewable linkages in terms of some general dynamical characterization.

10.4.1   *Non-renewable Resources Depletion*

Non-renewable resources are those naturally existing materials which have formed on time scales exceeding human lifetimes by order of magnitudes (e.g. minerals, petroleum, ...) and for which a natural regeneration term need not enter in their dynamical assessment. The determination of the location and quantity of such non-renewable resources generally represents a costly and time-consuming undertaking. In addition to the various land-use permits required, considerable planning of transport and markets need to be examined. Then, special devices have to be designed, assembled, and moved to the site and deployed. Among the most widely used exploration methods are direct boring and assessments of core specimen involving interpolation and statistical-probabilistic estimations for regions adjacent to the bore sites. Induced seismic shock wave methods are also often used, yielding indirect information about subsurface composition. The results from such resource explorations constitute proprietary information and are invariably closely guarded.

Consider then, as an example, that a specific fossil fuel resource has been identified and that its local *in-situ* quantity may be represented by $X(t)$ in some convenient units such as metric tons (10^3 kg), barrels (159 L), or energy (Joules, m^3, tons-of-coal equivalent, ...). With regeneration and replacement occurring at an insignificant rate, the $X(t)$ dynamic is therefore simply

$$\frac{dX}{dt} = -g_-, \tag{10.5}$$

where g_- is the extraction or withdrawal rate.

Two elementary autonomous extraction rates are of frequent interest:

(a) Constant quantity extraction rate: $g_-^q = \alpha_0$
(b) Constant fraction extraction rate: $g_-^f = \alpha_1 X$

By substitution, the time evolution of $X(t)$ can be as depicted in Fig. 10.3. As indicated, the constant quantity extraction rate leads to total resource

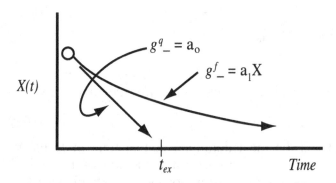

Fig. 10.3   Non-renewable resource $X(t)$ retention for the case of constant quantity extraction rate, $g_-^q = \alpha_0$, and constant fraction extraction rate, $g_-^f = \alpha_1 X$.

depletion at $\tau_{ex} = X(0)/\alpha_0$ while the constant fraction extraction rate approaches resource depletion asymptotically as $t \to \infty$.

The depiction Fig. 10.3 implies that $X(0)$ is known. In practice, however, it is very difficult — if not impossible — to specify the total quantity which could be extracted and this for two reasons:

(a) *Measurement*

All resource inventory assessments are based on sampled data and hence involve statistical uncertainty.

(b) *Extraction Cost*

The quantity extractable depends upon the cost of extraction; hence, at higher cost of extraction, even the less accessible lower concentration and smaller resource fields could be mined.

For this latter reason, resources may never be fully exhausted — they may however become prohibitively expensive to extract, thereby prompting alternative or substitutional developments.

In addition to the time variation of the extraction rate of a recoverable resource and cost of extractions, resource extraction interests must simultaneously consider other *down-stream* factors. A currently important case involves the resource of hydrocarbon fossil fuels. Some considerations relate directly to the primary exoergic chemical reactions and predominantly involve the rearrangement of hydrocarbon molecules of which the following is a useful example:

$$CH_4 + \begin{array}{c} fuel \\ contaminants \end{array} + (O_2, N_2) \to \begin{array}{c} energy \\ gain \end{array} + 2H_2O + CO_2 + \begin{array}{c} particulate \\ contaminant \\ NO_x, SO_2, \ldots \end{array}.$$

$$(10.6a)$$

A stoichiometric analysis yields

$$\begin{array}{c} 1\ mass\ unit\ of \\ CH_4\ fuel\ burned \end{array} \to \begin{array}{c} 2.75\ mass\ units\ of \\ CO_2\ gas\ produced \end{array} + \begin{array}{c} NO_x, SO_2\ and \\ other\ contaminants \end{array}.$$

$$(10.6b)$$

Since methane, CH_4, is the most dominant combustible molecule in hydrocarbon fuels, we obtain a simple rule: one ton of fossil fuel burned contributes nearly three tons of atmospheric contaminant CO_2, lesser quantities of NO_x, SO_2 and trace quantities of other elements and minerals; it is estimated that presently about $\sim 10^{12}$ kg of various types of contaminants are thus introduced into the atmosphere each year — a global average of $\sim$150 kg/person/year. And these contaminants are increasingly recognized for their interfering impact on the atmosphere including smog formation, greenhouse gas production, and terrestrial surface temperatures.

10.4.2 *Metal and Minerals*

In addition to fossil fuel, certain metals have become increasingly important. These include not only copper and tin which had been known since Prehistoric times and iron since Ancient times, but also modern metals such as aluminum and silicon, without which the aerospace and electronic industries as we know them would not be possible.

Whereas fossil fuels are extracted with relatively high concentrations, most metals recoverable from preferred ores possess a much lower natural enrichment. For example, iron, aluminum, manganese, and chromium are often found in the 50% enrichment range in some ores, while others are typically found at much lower concentrations: zinc ($\sim$3%), nickel ($\sim$1%), copper ($\sim$0.5%), and gold ($\sim$0.001%). Large quantities of ore material must therefore be moved and processed for the small concentrations of the desired elements.

Global metal extraction rates for each of the most widely used metals — iron, aluminum, copper and zinc — exceed $\sim$10^8 tons/year and continue to increase. Hence large piles of ore process waste have been and continue to accumulate at mine and mill sites. Further, since corrosive and toxic acids and solvents are continuously used in processing, these hazardous substances may also diffuse and accumulate so that separate holding tanks or settling ponds need to be used for purpose of sequestering. Additionally, many metal bearing ores contain toxic metals in small concentrations including arsenic, cadmium, and lead. Continuous monitoring of the surrounding air and ground water is an essential safety practice at sites of the front-end materials flow cycle.

Among the mineral resources of considerable interest are the phosphorus-based ores (phosphate rocks, fluorospar). Approximately 90% of such phosphors are destined for the synthetic fertilizer industry with the remainder for building materials such as gypsum. For applications similar to that of phosphates is potash, a potassium based mineral such as K_2O (potassium oxide) or K_2C_3 (potassium carbonate) and some calcium based minerals.

Silicon is an interesting case. While other metal and mineral extractions continue to increase, silicon which is most essential to the modern electronic industry, continues to be extracted at an approximately constant rate of only $\sim 10^3$ tons/year since continuing device miniaturization has compensated for its increasing uses. However, semiconductor grade silicon demands extreme purity requiring considerable energy and resulting in severe toxic wastes from various processing stages. Note also, Table 9.2, silicon is exceptionally plentiful, comprising 23% of the Earth's crust.

Chlorine-based minerals are components of many reactive compounds with its major source consisting of salt mines.

Finally, to be noted as a special case is the mining of nitrogen from the atmosphere for the production of fertilizer; this extraction is accomplished at relatively low cost by the Haber-Bosch process, Sec. 6.5.3.

A frequent occurrence in the mining industry is that extraction rates of ore may begin slowly during market build-up, then accelerate, and finally diminish for reasons of market saturation, excessive cost of extraction, or emergence of substitution products. There may exist one dominant extraction rate or, for reasons of market dynamics, periodic smaller rates of varying amplitude. With the dynamic specified by Eq. (10.5), one may evidently obtain the extraction rate from the resource depletion schedule. As indicated in Fig. 10.4, while the extraction rates $g_1(t)$ and $g_2(t)$ may

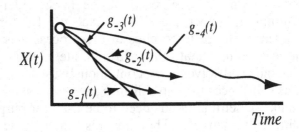

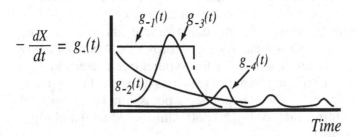

Fig. 10.4   Depiction of conceivable non-renewable resource depletion and extraction rates.

well be represented as implied in Fig. 10.3, the analytical representations of $g_3(t)$ and $g_4(t)$ as predictive functions are not normally possible and so follow from marketing considerations.

10.5 Renewable Resource Projections

In recent years engineers have had to become increasingly familiar with renewable resources since many engineering projects impact on wildlife and fish populations. As an initial point of clarification, a terminological distinction between renewable and non-renewable resources exists: what is named extraction rate for the latter is labelled harvesting for the former; further, fish, wildlife, livestock, forests, and other bio- and organic-entities are subject to natality and natural mortality.

The assessment of renewable resources involves important time variations with consequent implications on the interpretation of resource data. In particular, the dynamical features which add to the problematic and are of engineering interest are

(a) Mobility of wildlife and fish resources
(b) Age-dependent variations of renewable resources

A range of engineered devices now becomes important. For example, instrumentation which relate to aerial and surface mobility assessment of renewables (i.e. birds, game, ...) may require photographic and mobile telemetric devices, while fish population assessment in space and time requires suitably calibrated sonar and counting-categorizing of harvestable fish. In both cases, considerable statistical estimation and extrapolation is involved thereby introducing probabilistic considerations. In contrast, forest resources assessment involves periodic survey data including remote sensing and interpretation devices.

Projecting the general quantity of a particular renewable resource of interest begins with a dynamical description for a sufficiently large domain

$$\frac{dX}{dt} = r_+ - r_- - g_h, \tag{10.7}$$

where X is the time dependent total quantity of the resources, r_+ is the appropriate birth or regeneration rate, r_- is the corresponding death or decay rate, and g_h is the harvesting rate. Net migration is here taken to be zero implying therefore that inflow is balanced by outflow or that the domain of interest is essentially global.

For an animal resource population $X(t)$, (10.7) may be explicitly written as

$$\frac{dX}{dt} = a_+X - a_-X - g_h$$

$$= a_0X - g_h,$$

(10.8)

where the natality and mortality parameters α_+ and α_- are species specific. The harvest rate g_h is generally specified by regional custom and/or government policy directives.

Evidently the special case of $\alpha_0 < 0$ in Eq. (10.8) is of little interest since it represents monotonic depletion of the resource population; however for $\alpha_0 > 0$, an equilibrium bio-resource population can be identified by the condition $dX/dt = 0$ and gives

$$X^* = \frac{g_h}{a_0}.$$

(10.9)

As depicted in Fig. 10.5 and confirmed by an examination of its stability property, X^* is evidently an unstable population equilibrium.

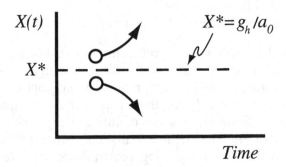

Fig. 10.5　Unstable equilibrium point and associated populations for a general renewable bio-resource subjected to a constant harvesting rate.

This unstable resource dynamic, associated with a constant harvest rate, possesses an interpretation of considerable relevance to resource management since the harvest rate g_h is often taken to be a human action control variable. Thus, should natural fluctuations lead to a population above the equilibrium X^* of Eq. (10.9), then — and without changing the harvest rate — the population will tend to increase. If the harvest rate is then increased to g_{h1} to define $X_1^* > X^*$ at that time then, the population trajectory will tend to decrease. Subsequently, in order to avoid extinction, the harvest rate must be reduced to g_{h2} to define $X_2^* < X^*$, a reversal in the population trend is thus initiated. This type of Bang-Bang control action is suggested in Fig. 10.6 and displays an oscillatory population variation with time; otherwise, either a population explosion or population extinction may occur.

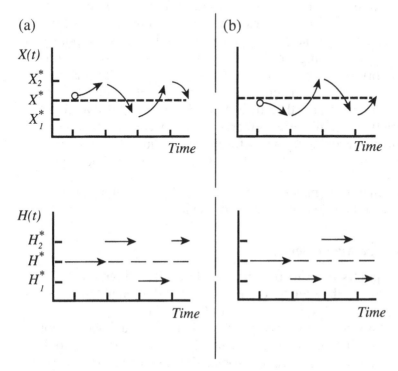

Fig. 10.6 Bang-Bang control of the harvest rate and fluctuation of the renewable bio-resource: (a) identifies an initial perturbation above X^* and (b) refers to an opposite initial perturbation below X^*.

An interesting economic optimization process may also be associated with the above Bang-Bang population dynamic. For the case of spontaneous fluctuations which increase the bio-resource population followed by an exponential growth in $X(t)$, the unit cost of harvesting is expected to decrease generating thereby additional demand and encouraging increased harvesting. Alternatively, should a random population decrease occur, then the unit cost of harvesting would increase, eventually leading to decreasing demand and a reduced harvesting activity allowing bio-resource recovery. Thus the variations of $X(t)$ tend to suggest a *feast-or-famine* resource time series unless exceptional management is possible.

10.6   Devices and the Engineer

It is important to appreciate that nature $N(t)$ and devices $D(t)$ form the primary interfaces for engineers $E(t)$ with the wider world. But for every generation of engineers, from the skilled stone-age engineers creating chipped tools, to clever and dexterous Roman engineers making a variety of ingenious artifacts and structures, and to Contemporary engineers making and using advanced software and hardware concepts, the same primal progression $N(t) \to E(t) \to D(t)$ has persisted. It was, however, in Ancient times that a societal interest $S(t)$ first became established and that in Medieval times a positive feedback from society to engineering emerged; this became extended in Modern times to also provide for a feedforward path thus finally yielding $N(t) \longrightarrow E(t) \longrightarrow D(t) \longrightarrow S(t)$. This latter concept has had a profound influence on how engineering developed and how thoughtful engineers came to view the role and function of devices.

10.6.1   *Knowledge About Devices*

Several general points need to be stressed about these ingenious and useful devices $D(t)$ as they relate to the engineer and to engineering.

First, as noted in Sec. 1.4, the diversity of Contemporary Engineering practice makes it necessary to employ a broad definition: a device is not only a hardware artifact, apparatus, structure, instrument, product, tool, appliance, etc., but also a relevant and associated software plan or process, such as strategy, maneuver, scheme, representation, improvization, etc. That is, a device is a human-made physical construction or cognitive abstraction which *engages*, *stimulates*, and *enables* its makers and users.

Second, engineers are obviously not the only ones involved with devices. Technicians in particular are trained in the installation, use, maintenance, and repair of devices. But engineers, more than any other group of specialists, are expected to know, to understand, and to communicate — at a high level of competence — the following aspects associated with devices related to their engineering discipline or to their area of technical expertise:

(a) What is the origin and past experience with a particular class of devices?
(b) Which device is the most appropriate for a specified application?
(c) Why should one choose a particular device from a conceivable set of devices?

(d) What are the consequences of the use and misuse of a particular device?
(e) How is a class of devices designed, manufactured, and maintained?
(f) What are the associated capital and maintenance costs?
(g) What are the risks and benefits of an existing or emerging device?
(h) What is the direction of continuing development of specific devices?

Answers to these questions reflect on the engineer's professionalism.

Third, devices have a history, or more specifically, they have a pedigree — a kind of genealogy. Devices have been conceived, designed, made, used, improved, and discarded for a million years. In the ensuing evolutionary process of adaptive variation and selective retention, their function and sophistication has undergone profound changes and affected personal practices of individuals and the status of communities. Thus, over time, some devices have been more important and more consequential than others; equality does not extend to devices.

Finally, devices do not just *do one thing* as suggested in the following examples. The invention and development of the bicycle did not just allow people to travel faster from point P_1 to point P_2; it did that but it also provided for a new status symbol, it required revised road-use patterns, it generated a new manufacturing and sales/service industry, it introduced a new form of recreation, etc. Fission reactors provided not only a superb submarine power source but it also enabled large-scale civilian electricity generation without the concurrent production of greenhouse gases, it provided for a new engineering discipline (Nuclear Engineering) and a new physics specialization (Reactor Physics), it stimulated the development of methods of nuclear medicine, it led to the need to manage a new type of process byproduct, it became an important factor in the politics of the Cold War, it introduced the process of industrial nondestructive testing by gamma-ray and neutron radiography, etc. The invention and subsequent development of charge-coupled devices not only improved observational astronomy but it also changed the way families record their important events, it allowed photography to become digital, it gave the military superb night vision capability, and pictures could be efficiently transmitted by e-mail. Indeed, every device does more than just serve one purpose or provide just one function.

10.6.2 *Origin of Devices*

To begin, presently identified distinct devices D_1, D_2, D_3, ..., D_i, ..., D_N represent a very large set. Patents filed over the past two hundred years number some 10^7. Considering that not all distinct devices are patented suggests that the set of all conceived patentable devices may well be in excess of 10^8, possibly approaching 10^9. Included in this set of inventions is the original telephone as well as the more than 3000 patents for catching mice and over 1000 patents for devices to be installed in wood/coal burning boilers to prevent chimney cinders from starting nearby fires. It is also sobering and deflating to note that only a very small fraction of patents have brought significant financial benefits to the inventor.

Consider the set of known devices D_i as a present state and now seek to explore their source. That is, what might be taken as the foundational origin of devices, noting that engineers are invariably viewed as professional inventors and innovators?

A frequently heard assertion is that someone happened to recognize a *need* for which a specific device could be conceived to meet this need[†]. The reverse also occurs sufficiently often: a device may be invented for which there existed no recognizable societal interest and a need is then manufactured by suitable advertising or market distortion. Indeed, it seems that some individuals are especially *gifted* to create new devices much as others have a musical, artistic, or literal talent. These two intellectual origins of device invention may be labelled as

(a) Response-to-need
(b) Hard-wired-compulsion.

In both cases, the inventor needs to dwell on three considerations:

(a) *Theory*
     The underlying physical basis of the device: physical laws, intrinsic processes, component compatibility, ...
(b) *Manufacture*
     The means of production of the device: design parameterization, resource availability, method of manufacture, cost of production, ...
(c) *Impact*
     The eventual impact of the device: efficiency, reliability, saleability, safety, novelty, social disruption, symbolic value, ...

[†]Note that there exist many widely recognized present needs which are not being met by inventions: nuclear fusion energy; earthquake predictors; tamper-proof locks; automobile engines with triple gasoline mileage and little change in power; water filters with 99.999% efficiency for common pollutants removal (and 10-year life span); etc.

Engineers are uniquely qualified in combining materials in innovative ways to meet specific operational features. This notion may be incorporated into the engineering progression as

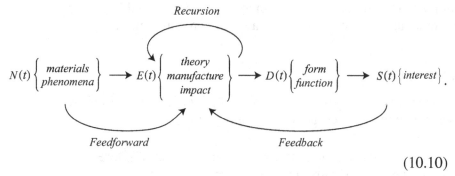

$$(10.10)$$

Note that feedforward tells the engineer what is possible, feedback conveys what may be wanted and is acceptable, and intellectual recursion seeks to identify the most appropriate specific parametric characterization of a device; it is these highly personal and unique intellectual processes which contribute to the eventual useful and ingenious device.

10.7　Stimulating Invention/Innovation

Invention and innovation are closely related concepts: the former is stimulated by some creative initiating idea — or set of ideas — and the latter refers to the ensuing analysis and actions in order to direct the initiating idea to commercial fruition. Invention may be some novel workable concept or demonstration for a device but it becomes patentable only if it meets specific legal criteria. In contrast, innovation is commonly taken to relate to marketing enhancement of some idea.

Another way to think of these related concepts is to recognize that invention (a noun) depends upon a creative impulse (i.e. one or several spikes in a person's brain signals) while innovation (a verb) is the process of commercialization (i.e. a set of adaptive activities). But, the process of innovation may often be considerably aided by some spike of creativity deserving, perhaps, to be considered as a separate invention. Though conceptually distinct, invention and innovation are evidently related.

The duality process of invention and innovation, (10.10), can be stimulated in engineering — but not dictated — by four identifiable cognitive artifacts and heuristic processes.

10.7.1 *Eureka Occurrence*

This label refers to the process of occasional organized thought focussed upon some technical goal of interest with the expectations that aspects of inference and conceptualization leading to invention and/or innovation might spontaneously appear. A little common wisdom should here be noted: spontaneous inspirations tend to favor the prepared mind.

10.7.2 *Creation Process*

A systematic process in support of invention or innovation is aided by the following sequential heterogeneous progression:

(a) Preparation: familiarization with the problem domain of interest
(b) Incubation: devoting nurturing thought to the problem
(c) Formulation: identification of tentative solutions
(d) Confirmation: revising and refining of proposed solutions

Note that this is an easily followed prescriptive process but invariably may require much iteration and the avoidance of rigid and narrow thinking — colloquially known as *ruts*.

10.7.3 *Brainstorming*

A suitably prepared group engages in a focussed discussion/analysis process under strict adherence to specified rules of interaction — positive thinking only, encouragement of far-out ideas, promotion of deeper/newer perspectives, identification of potential connections, no on-lookers or casual listeners, etc. The resultant set of ideas, concepts, and linkages are then variously sorted, screened, and subsequently circulated to serve the purpose of triggering further and particularly relevant ideas, useful connections, and important applications.

10.7.4 *Permutation Choices*

This process seeks to combine an essential set of relevant and analytically linked device parameters in various conceivable design configurations with bounded ranges of numerical variables; each of the many feasible solutions is then subjected to a specified comparative test of outcome performance measures.

A symbolic representation of the above stimulation processes is suggested by

$$E(t) \left\{ \begin{pmatrix} thought \\ actions \end{pmatrix} : \begin{pmatrix} invention \\ innovation \end{pmatrix} \right\} = E(t) \begin{cases} Eureka \\ Creation \\ Brainstorming \\ Permutation \end{cases}.$$

$$(10.11)$$

Once an invention or innovation has been identified and established, specific details of device patentability, design, manufacture, utilization, need, and marketing require consideration.

10.8   Intellectual Property

Consider a basic question: do engineers possess intellectual ownership — as distinct from physical possession — in a particular device? Indeed, some engineering innovations can be of profound significance deserving of special recognition to those who originated or contributed to these notions. Legal protection is generally recognized for four distinct intellectual categories associated with devices.

(a) *Invention of a Device*
    This identifies hardware objects, detailed drawings and blueprints, specified manufacturing or process control methodologies, software programs, and unique composites of materials which had not previously been known in the form presented and for the uses envisaged; the distinction of what is new and what is an obvious variation is often a critical issue and may need to be resolved by a judicial process. Just a novel suggestion or idea, is not — in itself — considered an invention.
(b) *Works of Authorship*
    Reports, technical papers, scientific manuscripts, encoded information, and similar recorded materials are included in this category.
(c) *In-House Know-How*
    This refers to proprietary information, confidential trade practices, and classified institutional techniques.
(d) *Symbols*
    By symbols, one identifies not only logos or graphical corporate identifiers, but also catchwords, slogans, and phrases.

Since the use or application of any of the above forms of intellectual property may benefit potential users, it is universally accepted that the originator or sponsoring stakeholder should receive some benefit. This notion of reward contributes to part of the positive form of feedback in the engineering progression

$$N(t) \longrightarrow E(t) \longrightarrow D(t) \longrightarrow S(t) \longrightarrow R(t),$$

Reward feedback

(10.12)

and is evidently intended to encourage the continuing production of valuable intellectual property.

Like the manufacturing of physical goods, the manufacturing of intellectual *goods* is of importance to society. There exist, four specific legal instruments each designed to address one of the above four intellectual considerations.

10.8.1 *Patents*

A patent may be issued to an individual, a group, or a corporation for a legally recognized invention. The patent holder then has exclusive rights over all aspects of its manufacture, distribution, and marketing for a period of 17 years. Patents may be bought and sold like physical assets. Some patent applications have recently become controversial and their enforcement protection problematic because of ownership claims on (a) software subroutines, (b) e-mail business methods, and (c) the human genotype. Also, special patent rights may be legally introduced; varying period depending upon type of invention are also under consideration.

10.8.2 *Copyrights*

Copyrights may be declared in writing by the author for an original work of authorship and is applicable to any form of reproduction: mechanical, electronic, magnetic, etc.; the symbol © needs to be clearly displayed. This right to control the making of copies is retained for the life of the author, plus 70 years by the legal estate. Copyrights may be transferred or assigned. The copying of small parts of copyright material for private use or citation and review purposes is not considered a violation of copyright.

10.8.3 *Trade Secrets*

Trade secrets relate to particular practices or information or means of production which is specific to a corporate interest of sustaining a competitive advantage. There exist no terms of expiry and the conditions of protection rest entirely on confidentiality. Violation of trade know-how confidentiality constitutes a legal liability.

10.8.4 *Trademarks*

A non-generic word or distinct symbol may be registered as a trademark for exclusive use by the registrant. Its terms are without limit, providing it does not become generic. The symbol ™ must be displayed in association with each use of this trademark symbol.

In summary, one may write

$$D(t)\{intellectual\ ownership\} = D(t) \begin{Bmatrix} patent \\ copyright \\ trade\ secret \\ trademark \end{Bmatrix}. \qquad (10.13)$$

Patentability, and that of *inventing around* patents, are frequently troublesome issues in engineering. This difficulty stems from problems of distinguishing what features are sufficiently different to warrant being labelled new and distinct, what has or has not been previously known or sufficiently understood, and what has or has not been revealed in the patent application. In general, structural, mechanical, and electronic devices tend to be more difficult to protect, whereas materials based on uniquely definable molecular structure — as in chemical or pharmaceutical products — are more easily protected. Many corporations employ or retain patent lawyers and patent specialists for such purposes.

Laws concerning retention of trade secrets and in-house know-how, especially as they relate to their exit with departing employees who then join a competitor, have in recent years been successfully sustained in the courts.

10.9 Engineering Bounds

An idealistic view of engineering holds that engineers are able to pursue the designing and making of devices with exceptional freedom. That is to say that few fundamental bounds are imposed on the primal

$$N(t) \rightarrow E(t) \rightarrow D(t). \tag{10.14}$$

In actual practice, numerous bounds invariably exist — some obvious and others very subtle. We consider here selected aspects of such bounds.

To begin with, we had already noted in Chapter 1 that the workings of engineering relate to three specific features of nature, $N(t)$:

(a) *Foundational Information*
Nature is informing about physical phenomena and processes which relate directly to engineered devices.
(b) *Basic Resources*
Nature constitutes a resource of materials for the making, operating, and maintaining of devices.
(c) *Self-Dynamics*
Nature sustains a complex dynamic consisting of organic and inorganic processes and entities which may be affected by engineered devices.

To these three features of nature we now add the further two considerations discussed in Sec. 9.7:

(d) *Transcendental Engagement*
Nature may induce a sublime sense of awe and delight about its composition, configuration, and processes.
(e) *Sense-of-Place Attachment*
Nature engenders a pragmatic sense of geographical identification with and expectations about some region of the human environment.

Further, we add two important so-called *schools of thought* of relevance to the theory and practice of engineering. In brief, they may be summarized as follows:

(a) *Scientific/Objective Rationalism*
This consideration has its origin in the Scientific Method. It claims that all aspects of the physical universe $N(t)$ are subject to natural laws providing thereby a rational and objective basis for the design and manufacture of all devices $D(t)$; personal preferences and ideological considerations are of little relevance.

(b) *Social/Subjective Construction*

This consideration is of more recent origin. It holds that contemporary culture — implicit in $S(t)$ — forms a matrix of social relations which have the effect of selectively imposing subtle criteria on acceptable productions and functions of all devices $D(t)$.

In terms of our primal engineering progression, these two schools of thought seek to impose the prominence of feedforward — that is scientific/objective rationalism — or prominence of feedback — that of social/subjective constructions — on the devices produced by engineers. These alternatives are suggested by the following graphical depictions:

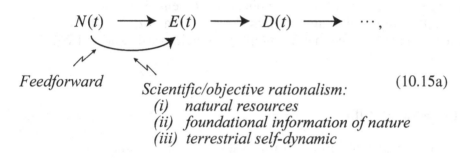

$$N(t) \longrightarrow E(t) \longrightarrow D(t) \longrightarrow \cdots,$$

Feedforward   *Scientific/objective rationalism:*   (10.15a)
     (i) natural resources
     (ii) foundational information of nature
     (iii) terrestrial self-dynamic

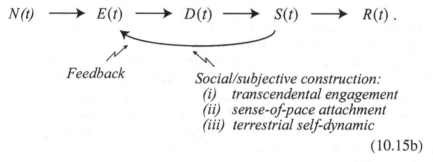

$$N(t) \longrightarrow E(t) \longrightarrow D(t) \longrightarrow S(t) \longrightarrow R(t).$$

  Feedback   *Social/subjective construction:*
      (i) transcendental engagement
      (ii) sense-of-pace attachment
      (iii) terrestrial self-dynamic

(10.15b)

Further, while scientific/objective rationalism and social/subjective construction are important boundary impositions on engineering activity, there still exists the additional consideration of employer directives for employee engineers

Employer: policy directive

(10.15c)

$$N(t) \longrightarrow E(t) \longrightarrow D(t) \longrightarrow S(t) \longrightarrow R(t),$$

which may involve corporate liability and financial considerations or restriction imposed by national/regional statutes. We add that for self-employed or consulting engineers, equivalent policy impositions or specific directives are provided by the client or customer.

By common educational experience and observed temperament, engineers tend to lean towards the scientific/objective rationalist perspective; in contrast, non-engineers tend to the social/subjective constructionist perspective. And therein lies a common dilemma for Contemporary Engineering: (a) continuing the objective conceptualization, design, development, manufacturing, maintenance, and ultimate disposal of devices without undue restrictions on the application of natural processes, (b) recognizing the legitimate societal/subjective socially constructed interests, and (c) further complicating the engineer's technical work are the employer's directives. Seeking a harmonization of the above three bounds has become a common implicit consideration by engineers.

10.10   Professional Ethics

The development and application of various devices is, by common experience, known to lead to a range of substantial societal consequences and it is widely recognized that some strictures are not only desirable but essential. Already during the Industrial Revolution, legislation was introduced abolishing child labor in factories and standards were established on industrial working conditions. In subsequent years, these restrictions were updated and other laws on workplace safety, minimum wage, job discrimination, working environments, and grounds for dismissal were enacted. More recently, legislation on standards for device safety, disposal of devices, use of devices in privacy invasion, and environmental impact of devices, have become law in many jurisdictions. Engineering licensing bodies also identify conditions which may require a process leading to the possible suspension of a licence to practice or to the imposition of penalties.

The idea of a *public good* and individual *moral responsibility* has become a common fulcrum for such consideration and this for good reason. For engineers these considerations are important because one may well identify numerous situations of responsibility in conduct and accountability where issues relate to an engineer's technical knowledge and professional obligations, and hence may lead to particular actions. The following list some illustrative examples about which an engineer may possess some *insider*'s knowledge with ethical implications:

(a) Device performance which does not meet published specifications
(b) Deliberate withholding of relevant technical information for fraudulent purposes
(c) Contract bidding influenced by illicit considerations
(d) Evident conflict-of-interest in technical and managerial decision making

⋮

Modern engineering has also seen the grass-roots emergence of a most proactive form of personal initiative: *whistleblowing*. This label follows from the common practice of a referee blowing a whistle when, on the basis of observations and understanding of the rules of the game, an infraction seems to have taken place. An engineer may similarly develop a conviction that an infraction of *good and fair engineering practice* has taken place or violation of an engineering related *public trust* has occurred. Actual forms of engineering whistleblowing generally fall between two extremes:

(a) *Lower Extreme*
An engineer has direct credible information which clearly shows an absence of good and fair engineering practice. This information is communicated to the supervisor following which corrective action is promptly implemented.
(b) *Upper Extreme*
In contrast to the above, an engineer communicates selected evidence alleging violation of a public trust together with an accusatory or inflammatory commentary to the public media; a rancorous public controversy then erupts.

Critical to the integrity of whistleblowing is the importance of verifiable evidence and clear violation of good and fair engineering practice; vague suspicions are insufficient. In recognition of the valuable public service possible by blowing the whistle, many jurisdictions have enacted legislation to protect whistleblowers from reprisals; nonetheless, such episodes can lead to considerable personal stress and career disruption for the whistleblower.

The fundamental and general issue then is the question of *how should one act in a situation* which possesses moral implications; that is, what principles or guidelines should be used? Indeed, even how one should think about such situations has occupied the thoughts of many. Already the Greek philosopher Aristotle (384–322 BCE) suggested that there exist certain universal virtues which are essential to a personal and community *good life* and these needed to be protected, preserved, and promoted. Then the rise of Humanistic thought of the late 1600s generated a new focus; for example, the 17th century English philosopher John Locke (1632–1704) taught that all humans are endowed with some *inalienable rights* and these could not be infringed upon by the actions of others. Another facet to this *rights* issue was introduced in the 18th century by the philosopher Immanuel Kant (1724–1804) of Germany who formulated the idea that every individual has a duty to act in a community-serving moral manner; his *categorical imperatives* specifies actions which would be generally acceptable and practised by others in the same situation. By the 19th century, utilitarian thought became of increasing importance and Jeremy Bentham (1748–1832), an English philosopher and jurist, suggested that when a dilemma emerged with moral implications, those actions which generated the *most good for most people* should be pursued. A subsequent perspective, enunciated by the 20th century Albert Schweitzer (1875–1965), German-French, was the notion of the primacy of *reverence for all life* as a guide for action.

Beginning in the 1960s, most professional societies developed Codes of Ethics. These Codes sought to establish guidelines of conduct for their members when confronted by ethical problems in connection with their engineering activities. In general, these Codes embody the following two principles:

(a) Members of a professional association of engineers accept the primacy of the principle that they possess a duty to (i) the public, (ii) their employer or client, (iii) their colleagues and associates and (iv) themselves.
(b) These duties require them to always act with fairness, fidelity, and integrity.

Such Codes of Ethics provide an itemization of specific guidelines for conduct and are frequently updated . Additionally, journals and newsletters of the profession report on particulars, such as guidance for whistle-blowing, legal liabilities, enforcement provisions, and specific cases of interest especially those involving highly visible device failures, public safety, and environmental protection.

The discussion of this last section in this chapter thus adds another dimension to the workings of the engineer. We incorporate this in the elaboration

$$N(t) \rightarrow E(t) \left\{ \begin{pmatrix} thought \\ actions \end{pmatrix} : \ ethical \ conduct \right\} \rightarrow D(t) \rightarrow \cdots$$

(10.16)

in the engineering connectivity.

10.11   To Think About

- What unique personal, intellectual, and manual skills or talents are typically expected of contemporary engineers and how do they differ from those of other professions?
- Project decisions often need to be made when incomplete or even ambiguous data are available. Consider the notion of resilient and adaptive outcomes under such input conditions.
- Could Eq. (10.7) be used to characterize commercially sustaining forestry management and stocked-lake fishing? Explore.

CHAPTER 11

Devices: Properties
and Functions

$$\cdots \rightarrow E(t) \rightarrow D(t) \rightarrow S(t) \rightarrow \cdots$$

11.1 Basics of Devices

The definition of a device, Sec. 1.3, includes both a material and cognitive characterization. Thus, the emergence of a device involves processes such as conceptualization, design, and manufacture so that one may think of a device as embodying considerations of

$$\cdots \rightarrow E(t) \rightarrow D(t) \begin{Bmatrix} \textit{matter} \searrow \\ \quad \textit{device} \\ \textit{process} \nearrow \end{Bmatrix} \rightarrow S(t) \rightarrow \cdots \quad (11.1)$$

and extends not only to organic/inorganic entities but also to creative/cognitive processes.

Devices, however, are more than just an integrated collection of detached and impersonal component objects and processes: devices are also *shapers of existence*. To appreciate this notion, one only needs to imagine life without refrigerators, cars, aircraft, television, thermometers, baseballs, turbines, bicycles, stethoscopes, electric power, row boats, wristwatches, elevators, keyboards, telephones, ... Or in more experiential terms, devices are the essential components of circumstances leading to delight as well as to agony: the lively banter around the warmth of a fireplace; the anguish of a serious train derailment; the pleasantly surprising long-distance telephone call; the mixed blessing of a recently completed nearby freeway interchange; the physical comfort due to air conditioning and mosquito spray; the apprehensions following a local processing plant explosion; ....

And, serving an essential role in these mixed experiences are the designers and makers of these devices — the engineers.

11.1.1 *Devices: Simplicity and Complexity in Space and Time*

Simple hardware items are homogeneous objects such as nails and paper clips, with their principal properties of material hardness and elasticity. But even these simple devices can be shown to possess a rich history reflecting on the creativity and persistence of inventive and innovative individuals in different locations and at different times.

Most devices invariably involve a complex range of engineering thought, actions, supportive tools, and interconnected processes. An automobile involves some 3000 assembly components, conceived, designed, developed, and improved over time, and manufactured in different locations but all scheduled to arrive at a point on the assembly line at a specified time. A space craft may involve perhaps 10^6 to 10^7 components, many of extreme specialization often developed under conditions of changing design criteria, expected to operate in an inadequately known environment, needing to perform functions which are continually evolving — and perform with extremely high reliability. A positron emission tomography device requires careful integration of nuclear, mechanical, thermal, and electronic processes all combining to provide a suitably detailed visual display of a biological entity.

It is interesting to note that many devices have emerged in time when the underlying physical principles were not adequately known: ships were built before the law of buoyancy had been formulated, mechanical weight-driven clocks were operating before the concepts of angular momentum and conservation laws were recognized, and the steam engine became a useful source of power before the science of thermodynamics emerged. Indeed, the evolution of devices has often stimulated the development of specific sciences.

Devices also diffuse in space and time. Construction of railroads spread from England to central Europe arriving about 1840, and continued to Portugal in the west and Russia in the east by about 1860. By analogy, devices seem to spread by diffusion much like a drop of milk in coffee but with a most significant difference: as they migrate, devices may also adapt in form and function and they may alter public attitudes. Thus, both the device and its environment may change simultaneously.

11.1.2 *Devices: Success, Failure, and Precariousness*

All devices are subjected to the possibility of success, failure, or precariousness — either prompt or variously delayed.

Bessemer steel production proved to be an unquestioned success to the continuing development of the railroad industry, the Bell telephone revolutionized communication, and the Model T established the remarkable effectiveness of assembly-line production.

Devices may fail while in use with spectacular and often catastrophic consequences: bridges collapse, dams fail, and aircrafts crash. Other devices fail to reach marketability in their developmental stages: sail-powered railroads went nowhere and the nuclear-powered aircraft never took off.

The supersonic transport (SST) airplane has long existed in a precarious state, the future of turbine powered trucks continues to be highly uncertain, and the continuing interest in special industrial applications of airships is still speculative.

Consideration of success, avoidance of failure, and elimination of precariousness of devices require considerable thought in design, manufacture, and operation; and for large device undertakings, a suitable strategy, political support, and public acceptance is generally of great importance.

11.1.3 *Devices: Users and Converters of Energy*

All devices require or convert energy — no exceptions.

We need to note first that energy is a most ubiquitous entity: it is absolutely essential to life and to existing human institutions. At its most basic level, energy possesses two primary origins:

(a) *Biological*

Energy is manifested in human and animal exertion and in the sustainment of organic functions.

(b) *Physical*

Energy is manifested in three physical contexts:

   (i) Matter-in-combustion: nuclear, atomic, and molecular reactions
   (ii) Matter-in-motion: environmental/gravitational consequences
   (iii) Matter-in-decay: spontaneous matter decomposition.

The widely used and very convenient electrical energy — AC or DC — is of secondary form having its most common origin in the combustion of hydrocarbon fuels, hydro/gravitational-to-electrical conversion, or heavy element fission; additional smaller but generally increasing contributions are provided by solar radiation, wind, tidal action, and geothermal sources. Also, thermal energy consists of nuclear motion, atomic motion, and molecular motion, all having their stimulation in some forms of reactions. Further, all forms of electromagnetic radiation are a secondary form of energy having their origin in charged particle motion. Finally, we note that solar energy is nuclear in origin.

The direct connection between devices and energy is absolute and extensive: devices require energy at all stages of production — mostly human exertion, thermal diffusion, and electrical stimulation — and they all require various forms of energy in their deployment. This latter requirement involves primary hydrocarbon fuels (e.g. for transportation, electric power generation, ...), electrical energy (e.g. for domestic appliances, process control, ...), and solar energy (e.g. for space heating, isolated instrumentation, ...). But most significantly, device usage has, with time, tended to involve a substantially decreasing requirement for human energy — with the obvious exception of devices specifically designed for fitness and recreational purposes.

An increasing consideration of Contemporary Engineering involves assured sources of primary energy, efficiency of energy conversion, and the matching of device function in relation to its energy requirements. Furthermore, all forms of primary-to-secondary energy conversion involve undesirable products or side effects requiring careful abatement considerations.

11.2   Pluralism of Device Functions

Planning for and designing of devices involves numerous considerations, including the possibility of misapplication as well as prospects of failure. Many engineers must have thought of an ideal device suitable only for a generally recognized desirable purpose: matches for lighting campfires but incapable of being used for arson; aircraft with flawless take-off, cruising, and landing but incapable of crashing; electrical/mechanical control devices which during a specified time operate with assured flawless performance; electronic communication devices which do not interrupt higher priority functions; etc. The world of devices has, however, evolved with exceptional pluralism, some of which may be characterized as follows.

11.2.1   *Device Disturbances*

Among societal disturbances caused by the appearance and use of devices, it is common to cite examples associated with adverse consequences; the following is a short listing:

(a) Changes in community practices
(b) Requirement for personal relocation
(c) Emergence of unforeseen commercial disadvantage
(d) Challenging established cultural practices and perceptions
(e) Increased noise pollution
⋮

Devices with a positive consequence can also be identified:

(a) Increased health and longevity
(b) Expanded protection of the natural habitat
(c) Improved life-style choices
(d) Enhanced employment opportunities
(e) Reduced environmental pollutant production,
⋮

Evidently, listings of the above type could be extended and specialized but five general groupings can be identified.

11.2.2 *Device Ambiguity*

The balance between advantage and disadvantage of devices may often introduce difficult questions even under normal operating conditions. For example, automobile air bags may well save lives but occasionally they are the cause of death; a brand of pesticide or herbicide may well be most effective in controlling a particular crop infestation but its appearance in the environment may also pose a potential carcinogenic hazard for other living species; software may provide superb radiation dosage control in medical therapy but an unrecognized programming glitch may lead to a misleading effect or even prove lethal.

11.2.3 *Device Misuse*

Users of devices do not always exercise good judgement or follow instructions. Thus, ladders and household stoves are essential but their common misuse contributes to frequent domestic personal injury; computer data banks may well provide for considerable administrative efficiency but unscrupulous operators may also engage in theft of data and invasion of privacy; the substitution of *nearly* equivalent replacement parts may lead to unintended adverse operational consequences.

11.2.4 *Device Inequity*

Advantages and disadvantages in the use of some devices may well involve separate communities. High speed air travel is a boon to frequent travellers but the associated aircraft noise may represent a heavy burden for those living near airport runways; a communications tower may well contribute to enhanced social and business contact but adversely affect the resale value of adjacent residences; the discharge of some process effluent into a river may well lower the cost of some manufacturing operation but it may also pose hazards to downstream users of water.

11.2.5 *Device Failure*

Devices are invariably also designed to minimize uncertainty in performance of some activity; that is, they are expected to function in a particular way in support of a particular purpose — but sometimes they fail. Electrical equipment provides important services but it may also cause severe fires; a dam may well provide for flood control but it may also fail causing downstream devastation; a car is a most convenient form of transportation but tire blow-out may lead to a serious accident.

Features of device pluralism invariably create problems of trade-off so that one may write

$$D(t) \left\{ \begin{matrix} pluralistic \\ properties \end{matrix} \right\} = D(t)\{trade\text{-}off\} = D(t) \left\{ \begin{matrix} disturbance \\ ambiguity \\ misuse \\ failure \\ inequity \end{matrix} \right\} \quad (11.2)$$

as a detail of device invention and innovation.

11.3 In-Use Failure

Engineered hardware devices may be designed to perform specific on-off or graduated control functions for purposes such as generating power, manufacturing products, transforming matter, manipulating data, and other processes. For example, if $X(t)$ is taken to be the number of devices of a particular kind all operating in a common domain, normal device operation then represents a stationary state of $X^*(t)$ or its average $\bar{X}$:

$$\left. \frac{dX}{dt} \right|_{X^*} = 0 \quad or \quad \left. \frac{dX}{dt} \right|_{\bar{X}} = 0. \quad (11.3)$$

Controlled deviations during some Δt are non-equilibrium start-up and shut-down processes.

Of interest now is the abnormal performances associated with device failure. For some general failure, the associated dynamic about some failure time coordinate t_x is evidently

$$\left.\frac{dX}{dt}\right|_{t_x} \neq 0. \tag{11.4}$$

While the details of the processes for $t > t_x$ may be most important in a reconstruction of failure or accident dynamics, a common interest is in the determination of t_x or the operating time interval $(t_x - t_0)$ during which a probability of failure might be specified. Two classifications of in-use failure are of particular interest because of their wide applicability:

(a) Cycle-of-life failure
(b) Severity-of-impact failure

Cycle-of-life failure is associated with the common notion of wear and tear. Since it relates directly to an intrinsic property of material degradation, the rate of failure depends upon the time period of use or intensity of use. Switches, rotating machinery, and pulsing devices commonly fail by this cycle-of-life failure mode. Simply put, cycle-of-life failure means device *wear-out* or *burn-out*.

In contrast, severity-of-impact failure is attributable to extremes of shock-type causes, the effect of which exceeds some specific tolerance design parameter of the device. Examples of such severity-of-impact device failure are building collapse due to severe earthquakes, dam failure due to high floods, and mishandling of shock-sensitive instruments.

In addition to the above two very common failure modes, there exist other causes of failure such as operator error, computer software deficiencies, sabotage, and the sporadic actions of cosmic rays or radioactive decay in adversely affecting low-current solid-state circuitry.

11.3.1 *Cycle-of-Life Failure*

Cycle-of-life failure of a device is commonly examined within the context of a failure in a large population of sufficiently similar devices. It is a dynamic process and may be considered to be akin to the death of a biological entity from a specific species population. In each case the population of operating devices has decreased in number. Device failure for the case of cycle-of-life failure is subjected to the basic survival dynamics of populations in general and a fractional decrease in a population can be directly related to a probability of failure for any of the identical devices.

The fundamental notion of an initial population of identical devices $X(0)$ all of which are either placed in service or subjected to simulated life-cycle testing at $t_0 = 0$, is a most suitable conceptual starting point. A typical time variation of the remaining number of functioning objects $X(t)$ is suggested in Fig. 11.1; this number is, appropriately, labelled the survival population or survival function $X_s(t)$ and is characterized by three distinct time periods:

(a) $0 \le t < t_1$:   early-life deficiency,   (11.5a)

(b) $t_1 \le t < t_2$: mid-life consistency,   (11.5b)

(c) $t_2 \le t$:   late-life expiry.   (11.5c)

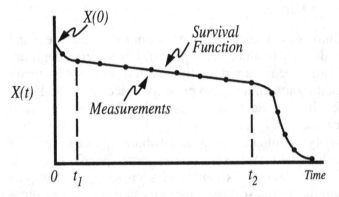

Fig. 11.1   Typical observed survival function $X_s(t)$ associated with an initial large set $X(0)$ of devices subjected to simulated cycle-of-life failure tests.

With $X_s(t)$ as the survival population, the companion cycle-of-life failed population $X_f(t)$ is evidently

$$X_s(t) + X_f(t) = X(0). \qquad (11.6)$$

The failure rate of the devices, r_f, is given by the rate equation for the cycle-of-life failure process

$$\frac{dX_f}{dt} = r_f. \qquad (11.7)$$

Possible functions $X_s(t)$, $X_f(t)$ and $r_f(t)$ are graphically suggested in Fig. 11.2 for typical cases.

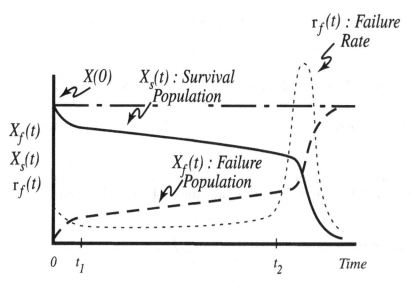

Fig. 11.2   Schematic depictions of conceivable device survival population $X_s(t)$, failed population $X_f(t)$, and the failure rate $r_f(t)$. We add that similar distributions apply to human populations.

A procedure which provides a useful tool for analysis follows by first integrating Eq. (11.7) to give

$$X_f(t_i) = \int_0^{t_i} r_f(t)\, dt, \tag{11.8a}$$

since $X_f(0) = 0$. Upon dividing by the constant initial population number $X(0)$ we write

$$\frac{X_f(t_i)}{X(0)} = \int_0^{t_i} \frac{r_f(t)}{X(0)}\, dt$$

$$= \int_0^{t_i} p_f(t)\, dt, \tag{11.8b}$$

with the function $p_f(t)$ evidently subjected to the normalization

$$\int_0^{\infty} p_f(t)\, dt = 1. \tag{11.9a}$$

The significance of $p_f(t)$ is due to the following interpretation of the ratio (11.8b):

$$\frac{X_f(t_i)}{X(0)} = \textit{fraction of the initial } X(0) \textit{ devices which failed by time } t_i$$

$$= \int_0^{t_i} p_f(t)\, dt. \tag{11.9b}$$

Since this qualitative statement holds for all t_i and (11.8a) is monotonically increasing with time, one may introduce the probability statement

$$Pr(\textit{device failure in the time interval 0 to } t_i) = \int_0^{t_i} p_f(t)\, dt. \tag{11.10}$$

It is also evident that

$$p_f(t)\, dt = \textit{probability that the device will fail in the interval dt about t}$$
$$\textit{(i.e. it has survived until time t and then fails during dt).}$$

This function $p_f(t)$ is commonly called the *probability density* failure function due to cycle-of-life failure. An indication of the general utility of a probability density failure function is that it allows the calculation of numerous failure probabilities of common interest:

$$Pr(\textit{device failure during time interval } t_i \textit{ to } t_j) = \int_{t_i}^{t_j} p_f(t)\, dt, \tag{11.11a}$$

$$Pr(\textit{device failure after time } t_x) = \int_{t_x}^{\infty} p_f(t)\, dt. \tag{11.11b}$$

The expression *cumulative probability* for the indicated time interval is often used to describe expressions such as Eqs. (11.10) and (11.11).

With the population of the original devices depleting with time, Fig. 11.1, it is of frequent interest to establish some characteristic measure of service of the device given that the probability density for failure, $p_f(t)$, is known. Longevity of service is clearly an important parameter and the statistically expected mean operational life-time of any one device follows by formal definition:

$$\tau = \int_0^\infty t p_f(t) \, dt. \qquad (11.12)$$

The reliability of a device up to time t, $Rel(t)$, is also a most useful concept and is defined as the probability of non-failure up to time t computed from

$$Rel(t) = 1 - \int_0^t p_f(t) \, dt. \qquad (11.13)$$

Finally, for a significant period of time, $p_f(t)$ may be well approximated by a constant $p_f(t) \to \lambda_f$; for self-evident reasons this is often called the *bathtub* approximation, Fig. 11.2. If this approximation can be considered to be acceptable for the time interval for $t = 0$ to $t = t_x$, then device reliability, (11.13), takes on a particularly simple form

$$Rel(t_x) = 1 - \int_0^{t_x} \lambda_f \, dt$$

$$= 1 - \lambda_f t_x. \qquad (11.14)$$

A sufficiently small λ_f is clearly desirable for any device.

Equipment reliability considerations have led to the establishment of two common device replacement strategies. One simply involves *replacement-upon-failure* and is commonly practised with devices where failure is of little consequence (e.g. household light bulbs, automobile batteries, etc.). The other strategy involves *replacement-after-a-usage-time t_u*; aircraft tires and transportation corridor signal controllers are examples of the latter category because failure could lead to very serious consequences. Note however that the determination of t_u may now be important, typically determined by some acceptably small probability of device *failure* during its use.

11.3.2 Severity-of-Impact Failure

Severity-of-impact failure involves external causes which have the effect of exceeding a limiting operational parameter of the device. This may, as typical examples, involve hurricane winds exceeding a building's design wind resistance, temperature above or below which motors will fail to function, snow accumulation on a roof exceeding the beam carrying capacity, ice storm conditions leading to transmission power line failure, and numerous others.

Two properties of external causes are of dominant importance. One is that the magnitude of the causative phenomena often represents a statistical variate and can only be anticipated in terms of probability considerations; the other is that such phenomena may possess a seasonal frequency pattern. For both of these reasons, a historical — or otherwise validly reconstructed[†] — record of sufficient length is needed in order to provide a basis for the characterization of the causative phenomena. A severity-of-impact failure analysis may well be illustrated using river floods as an example.

Consider a traverse at a particular site on a river which is near some proposed construction project of interest; this project could be a dam, a bridge, a pier, or some building, the destruction of which is taken to occur when the transverse-area integrated mass flow rate of water exceeds a quantity Q_x in units of m³/s.

Next consider the unique maximum natural flow rate Q_i in a conceivable season of interest, Fig. 11.3. The sequence of such maxima from each season yields a time series of maxima, Fig. 11.4.

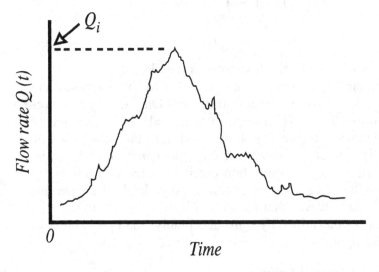

Fig. 11.3   Conceivable natural seasonal mass flow rate $Q(t)$ at a river site of interest.

The next step is to deduce from this sample-history its frequency distribution of floods and associated probability density function for the population of floods, $p(Q_i)$; the determination of this density function invariably involves subtle plausibility considerations based not only on

---

[†]For example, by correlation analysis involving related phenomena.

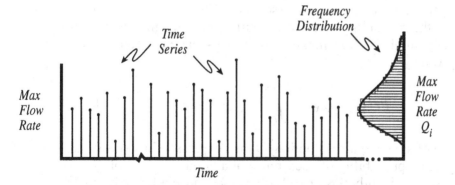

Fig. 11.4   Example of a time series of maximum annual flow rates showing time series of seasonal floods and their associated frequency distribution.

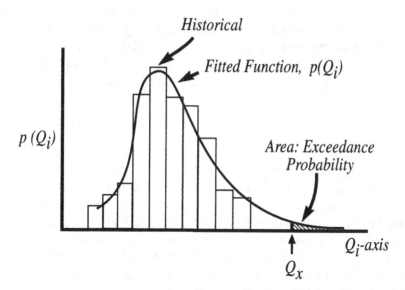

Fig. 11.5   Probability density $p(Q_i)$ fitted to a hypothetical historical frequency distribution. A design flood Q_x and its exceedance probability are indicated.

goodness-of-fit of $p(Q_i)$ to the historic record but also upon the underlying derivational basis of the function. Figure 11.5 displays a conceivable probability density function.

Figure 11.5 also suggests a possible design value for flood protection, here shown as Q_x. An important point to recognize is that relevant probability density functions are *right-side* unbounded so that a maximum possible flood magnitude cannot be specified; hence, one may only assert, to a given level of confidence associated with the determination of $p(Q_x)$, what the probability of a flood occurring with magnitude greater or less than Q_x might be. Specifically, a flood-per-year exceedance probability is given by

$$Pr(Q > Q_x) = \int_{Q_x}^{\infty} p(Q)\, dQ. \qquad (11.15)$$

The issue of safety against flooding is thus expressed in terms of an acceptably small probability of occurrence of a flood in excess of Q_x in any one season. Evidently this probability can be made arbitrarily small by choosing a sufficiently large Q_x — and accordingly adding to the cost of protection.

Another useful concept can be introduced by the following self-evident analogy: if floods occur on an annual basis — as for example snow melt floods — then if a given flood exceeding, say Q_x, has a 10% probability of occurrence in any one year, such a flood will occur on average once every 10 years. This introduces the concept of *return period* $\tau(Q_x)$ for the flood of magnitude Q_x and is evidently defined by

$$\tau(Q_x) = \frac{1}{\int_{Q_x}^{\infty} p(Q_i)\, dQ_i}. \qquad (11.16)$$

Note an interesting and perhaps unexpected conclusion: since the probability density function $p(Q_i)$ fitted to a historic sample record is unbounded for $Q_i \to \infty$, *absolute protection* for a severity-of-impact failure involving seasonal extremes is not — by mathematical/statistical reasoning — possible; an event more severe than any on record is always expected to occur, albeit with a decreasing probability. An acceptable probabilistic unlikelihood of an event thus becomes an engineering design criteria arrived by societal safety considerations.

11.3.3 *Event Trees*

It often happens that, while individual events occurring in isolation may be of little consequence but if they are part of a critical chain of events then, conceivably, very substantial damage may result. Such cases may be analyzed by an itemization of connected but independent probability events placed in their proper serial and parallel order thereby providing a clear identification of critical and noncritical sequences of events; the term *event-tree* is commonly assigned for such characterizations.

Consider the establishment of an event-tree for the case of a marine tanker entering a harbor and, depending upon certain events, it may or may not lead to spillage of crude, Fig. 11.6. As shown in this figure, the probability of spilling crude in the harbor for each tanker entry is evidently given by

$$Pr(spill) = {}_1P_1 \times {}_2P_1 \times {}_3P_1, \tag{11.17a}$$

while the converse probability is

$$Pr(no\ spill) = {}_1P_2 + ({}_1P_1 \times {}_2P_2) + ({}_1P_1 \times {}_2P_1 \times {}_3P_2). \tag{11.17b}$$

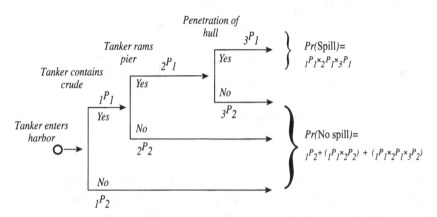

Fig. 11.6   Illustration showing an event-tree associated with a tanker entering a harbor.

The likelihood that changes in tanker and harbor design — as well as operational practice — will occur with time, means that the probabilities of events will also change with time. Hence, under such conditions event-tree probabilities are nonautonomous and such assessments have to be reviewed as necessary. Note also that since it is a product of probabilities which is the determining factor, an optimization of various hardware or procedural components may be undertaken to establish which probability components changes should be sought given limited resources.

11.4   Human Error

One of the most troublesome engineering projections relates to human actions associated with device operation. Notwithstanding the sayings that *to err is human* and *human brilliance is as unpredictable as human stupidity*, some organizational insight can be established and be of practical interest in engineering. Symbolically, this kind of human error problematic can be suggested by additional components to the engineering connectivity, Fig. 11.7.

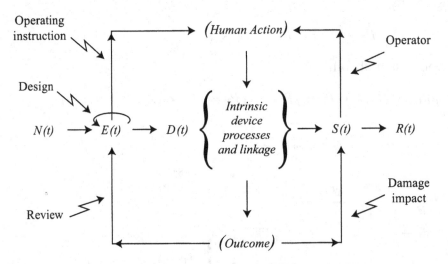

Fig. 11.7   Linkage schematic associated with human actions on devices.

An important point to recognize is that a device may possess some intrinsic linkages which results in specific dynamic behavior when triggered by particular human actions. To approach this subject consider a progression of human actions $Y_i(t)$ but with one action, say $Y_{error}(t_x)$ of particular adverse consequence:

$$Y_i(t_1) \to Y_j(t_2) \to Y_k(t_3) \to \cdots \to Y_{error}(t_x) \to \cdots \qquad (11.18a)$$

Additionally, adverse outcomes may come about because of a sequence of seemingly minor or innocuous errors:

$$Y_i(t_1) \to Y_j(t_2) \to Y_k(t_3) \to \cdots \to Y_{error1}(t_{x1}) \to$$
$$Y_{error2}(t_{x2}) \to Y_{error3}(t_{x3}) \to \cdots \qquad (11.18b)$$

Of the very large set of conceivable human error actions and their consequences, one may identify three dominant classifications.

11.4.1   *Minor Errors*

This refers to the numerous small human errors of moderate consequences. They may be associated with inattention, fatigue, distraction, ineptitude, etc., and involves little more than inconveniences and low cost consequences.

11.4.2 Transparent Errors

A most important category encompasses consequential human errors which have a clear and evident cause. Several classes in this category can well be described by examples.

(a) $Y_{error}(t_x) = $ *Ignorance*

A welder is working in close proximity to other equipment including — but unknown to this worker — an improperly labelled canister containing a flammable liquid; a spark from the torch causes an explosion.

(b) $Y_{error}(t_x) = $ *Faulty Hypothesis*

A locomotive operator discovers a faulty signal warning light and chooses to become especially vigilant until he reaches his final destination; the operators assumption that his senses could perform the same hazard-assessment functions as the signal warning light, proved to be in error and a serious derailment occurred.

(c) $Y_{error}(t_x) = $ *Rare Interference Effects*

The captain of a commercial aircraft decides to proceed with landing during a heavy rain and gusty wind conditions. A last minute very heavy downpour and severe wind shear forces an angled and imbalanced hard landing and slippage off the runway. It is evident that the captain failed to give adequate consideration to the combined effect of the adverse natural occurrences.

While device redesign may be important for such cases, specialized operator training is invariably also necessary.

11.4.3 Complex-Linkage Errors

Complex-linkage errors identify severe device failures involving multiple and interactive effects including the critical role of human actions in the process. Reconstruction of the precise dynamical sequence of events is often very difficult since critical device components may have become damaged. Further, for technical reasons and associated legal-liability consequences, the causes might not be agreed upon by the affected parties for a considerable time — perhaps never.

Such complex-linkage errors are especially important for engineers because some underlying circumstances may well be related to factors such as the following:

(a) *Haste*
   Excessive rush in design in order to meet externally imposed
   deadlines.
(b) *Neglect*
   Important tests and experiments may, for a variety of reasons, have
   been cancelled.
(c) *Tunnel-Vision*
   An excessively narrow design and operational perspective may have
   been adopted.
(d) *Diagnostics*
   An installation may have been provided with inadequate sensing
   instruments.
(e) *Bewilderment*
   Operators of complex devices may lack proper insight to interpret
   critical precursors.

The following lists some examples of device failures involving multiple design and operational errors.

(a) *Exxon Valdez Spill, Alaska (1989)*
   Oil tanker grounded; 10^8 L crude spills

   Y_{error1}: tanker deviated from marked channel
   Y_{error2}: person-in-charge failed to take corrective action

(b) *Capsizing Ferry, Zeebrugge Harbor, The Netherlands (1982)*
   Two-deck car ferry capsizes; 189 fatalities and $\sim$150 injured

   Y_{error1}: bow loading door left open during departure
   Y_{error2}: improper trimming of bow ballast
   Y_{error3}: life jackets kept in locked cabinets

(c) *Toxic Gas Release, Bhopal, India (1984)*
   Pesticide plant releases poison gas; $\sim$10,000 fatalities, $\sim$100,000
   injured

   Y_{error1}: gas detectors not operational
   Y_{error2}: safety bypass system disabled
   Y_{error3}: improper mixing of chemicals

(d) *Chernobyl Nuclear Reactor Explosion, Kiev, USSR (1986)*
   Electric power producing nuclear reactor experiences a runaway exoergic chain reaction; 32 immediate fatalities and widespread radiation
   exposure.

   Y_{error1}: disconnection of safety system for rundown test

Y_{error2}: rundown operation transferred core operating regime into a reactivity instability domain

Y_{error3}: excessive manual overcompensation discharged coolant thereby establishing excessive power density

Human performance projections, according to Eq. (11.18), requires the identification of human error actions Y_{error} associated with a complex chaining in device features and operation. In this context, the complex-linkage error discussion suggests a profound consideration for the engineer:

> Does the conceivable intrinsic dynamic of a device possess particular design and operation features which might — under credible isolated or interactive circumstances — involve a significant mismatch with the intellectual capacity or manual actions of the operator or decision maker and thereby lead to potentially severe consequences?

Thus, the projection of human performance leads to fundamental issues of human thought and actions, interacting with the dynamic properties of a device and associated environment in which the consequences might be played out. Increasingly, such considerations are being recognized as most important to the establishment of a thorough safety culture in device design, manufacture, and operation.

The point is sometimes made that the evolution of human decision making capability has not kept pace with the operational requirements of advanced engineered devices.

11.5 Forecasting

For many purposes in the practice of engineering, it is necessary that systematic methods and techniques be employed which lend themselves to quantitative device-related forecasting.[†] Determining the variation with time — for example that of efficiency, impact, sales, production costs, uses, etc. — is a common engineering interest. A number of forecasting techniques have been explored over the years of which the following have been found particularly useful for selected engineering applications.

[†]Note the important distinction between forecasting and prediction: forecasting involves analysis of various relevant factors while prediction is commonly taken to be simply an intuitive statement about the future. For example, weather is forecast, not predicted.

11.5.1 *Trend Analysis*

Of all forecasting techniques, Trend Analysis is the simplest for it involves only the extrapolation of one or several relevant measurable variables with time. Time variations of monthly sales, market fractions, production rates, etc., are particularly relevant. The basis for this technique of forecasting rests on the notion of expected smooth changes of all relevant parameters and processes; business-as-usual is a common description for such assumptions.

11.5.2 *Scenario Planning*

This technique involves the formulation of various conceivable evolutions of relevant exogenous variables based on the identification of an adequate and realistic range of causative developments; particular events and estimates of their probability of occurrence also need to be specified. This method is commonly employed in global resource and investment planning and may involve probability estimations of events such as specific political actions, social unrest, changing regulatory legislation, environmental disturbances, emerging community interests, and others. Multinational corporations frequently use this approach in large-scale industrial development planning.

11.5.3 *Reasoning by Analogy*

The basis for this approach is to identify relevant situations in different locations, or at different times, for which some similarities are sufficient to suggest transferable characterizations. At a more subtle level of analysis, one may also reason by analogy on different space and time scales involving possibly different phenomena. For example, the time-dependent acceptance of a new device in a region may be comparable to physical diffusions of some material agent in a diffusive medium. Characterizations such as impact-response, mean-time-to-adoption, and the time-until-replacement might then be estimated.

11.5.4 *Delphi[†] Technique*

Delphi forecasting involves the polling of experts. Then, the polling results are subjected to a clustering and frequency distribution analysis

---

[†]Delphi is the name of an ancient town in Greece which was the home of the Oracle of the Greek god Apollo.

and submitted for a revised forecast to the same group of experts but now expanded to including new participants. This technique is widely used in the analysis of new device development particularly if strategic research and development advances need yet to be clarified.

11.5.5 Dynamical Analysis

This forecasting methodology is based on the notion that it is often possible to mathematically specify the form of particular rates of change which affect some variable $X(t)$ of interest; indeed, such information is often established from historical data. Typically, relevant rate equations or dynamical representations are then of the form

$$\frac{\Delta X}{\Delta t} \rightarrow \frac{dX}{dt} = \sum_i \left(\pm \frac{dX}{dt} \right)_i , \qquad (11.19)$$

for which commonly encountered influence processes such as interactions, feedback, regeneration, catalysis, migrations, etc., can all be incorporated as specific rate terms $(dX/dt)_i$. This approach has in recent years proven to be insightful when combined with the graphical interpretation of differential equations. Note that the formulation (11.19) implies a prioritization by the analyst since the specific contributing rate terms chosen are invariably critical to the product of interest. At a more foundational cognitive level, this approach implies that the human mind is better able to specify particular rates of change $(dX/dt)_i$ as a function of time than to directly identify a reliable time projection of $X(t)$.

11.5.6 Complex Dynamics

The preceding method of dynamical analysis is extended by inclusion of the simultaneous evolution of the set of interdependent variables $X_1(t), X_2(t), X_3(t), \ldots$, and which are solutions of the coupled autonomous dynamical description of the nonlinear form

$$\frac{dX_i}{dt} \sim R_\pm^i + \sum_j a_j^i X_j + \sum_k \sum_l b_{kl}^i X_k X_l + \cdots \qquad (11.20)$$

$$i = 1, 2, 3, \ldots; \quad a_j \text{ and } b_{kl} \text{ are parameters.}$$

Solution analyses leading to dimensional reductions may identify unique patterns or *signatures* of the underlying phenomena. A range of applications of this approach — not only in science and engineering but also in economic market analysis — have in recent years been explored.

The preceding lends itself to a compact summary description of forecasting in engineering:

$$E(t)\left\{\begin{pmatrix} thought \\ action \end{pmatrix} : forecasting\right\} = E(t)\left\{\begin{array}{l} Trend\ Analysis \\ Scenario\ Planning \\ Reasoning\ by\ Analogy \\ Delphi\ Technique \\ Dynamical\ Analysis \\ Complex\ Dynamics \end{array}\right\}.$$

$$(11.21)$$

Of the above six techniques, Trend Analysis is evidently the simplest and most often used but it may also be most restrictive and unreliable especially if the variables of interest are carelessly chosen or if the near-term is expected to be sufficiently sporadic. Scenario Planning, Reasoning by Analogy, and the Delphi Technique are powerful tools of analysis but are critically dependent upon the experience of those undertaking such analyses. Dynamical Analysis has in recent years received considerable attention especially as a heuristic tool and has opened a number of new directions. Complex Dynamics, as evidenced from a comparison of (11.19) with (11.20), is a natural extension of Dynamical Analysis which makes increasing demands on the identification of suitable variables and parameter domains of particular relevance and associated geometric

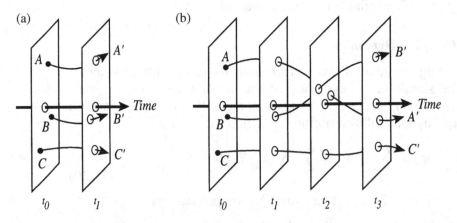

Fig. 11.8   Comparison between (a) short-term and (b) long-term projections in time. For the former, linear methods of extrapolation are often adequate but for the latter it is invariably necessary to use nonlinear methods which incorporate careful choices of interactive considerations.

representations. Experienced analysts generally seek to employ more than one method of forecasting.

This discussion of projections in time itemizes particular techniques of relevance to progressions in time. A self-evident rule is that for sufficiently long-term projections, complex nonlinear rate expressions — that is processes which reflect a variety of complex interactions — may need to be included; a graphical depiction of such linear and nonlinear projections is shown in Fig. 11.8.

It may often be of considerable interest to explore some general, and perhaps philosophical, questions about the continuing evolution of devices. For example, to what extent can one expect future devices to develop towards higher speed (e.g. ships, aircraft), smaller dimensions (e.g. calculators, telephones, ...), lower weight (e.g. cars, bicycles), and lesser cost (e.g. computers, appliances)?

11.6 To Think About

- Why are elevators and escalators so remarkably safe? Could the underlying principles be applied to other risky devices?
- Explore the meaning and applications of Human Factor Engineering.
- Engineers are often expected to identify the one firm projection to a given dynamical formulation. Discuss the possibility of robust engineered devices able to meet various rational projections simultaneously.

CHAPTER 12

Society: Involvement and Ramifications

$$\cdots \to D(t) \to S(t) \to R(t)$$

12.1 Societal Interest

As first noted in Sec. 3.7, the societal interest $S(t)$ in the core engineering connectivity progression $N(t) \to E(t) \to D(t) \to S(t) \to R(t)$ is primarily grounded upon the triad of economic, political, and religious considerations:

(a) *Economics*
   Considerations of subsistence, labor, trade, exchange, employment, commerce, production of a surplus, generation of wealth, etc.
(b) *Politics*
   Considerations of authority, governance, regionalization, security, border regulations, administration of justice, etc.
(c) *Religion*
   Considerations of spirituality, morality, faith, liturgy, piety, sacred places, communal rituals, etc.

And, quite clearly and most importantly, engineered devices may directly and indirectly relate to all aspects of this societal triadic interest:

$$N(t) \to E(t) \to D(t) \begin{Bmatrix} ingenious \\ useful \end{Bmatrix} \to S(t) \begin{Bmatrix} economics \\ politics \\ religion \end{Bmatrix} \to R(t).$$

$$(12.1)$$

Societal demonstration and expression of its interest in devices, however, is rarely without qualification or ambiguity. At a general level, society invariably recognizes advantages and disadvantages with the appearances and use of devices; at another level, it recognizes issues of risk and safety. One organized depiction of such a dilemma is possible by suggesting acceptance, rejection, or ambivalence as domains on a Cartesian plane characterized by some integrated measure of opportunity and danger as coordinate axes, Fig. 12.1. Consequent community choices may be readily possible if a preponderance of views or tendencies are found in the two domains of high-danger/low-opportunities or low-danger/high-opportunity; otherwise, polarization may develop resulting in a possibly problematic conflict. This conflict may be aggravated because device choices may relate directly to particular agenda characterizing special interest groups in a community. Indeed, it is not uncommon for particular groups to engage in extensive promotional support or opposition on device rejection or adoption.

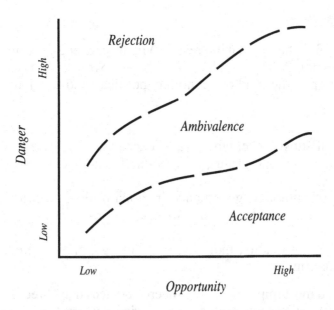

Fig. 12.1 Domain characterization associated with societal response to some device $D(t)$.

12.2 Risk and Safety

Some clarification on device acceptance or rejection may be possible by a quantitative assessment and qualitative evaluation associated with the device or set of devices of particular interest.

12.2.1 *Quantitative Assessment*

Engineers may contribute to a clarification of device considerations by a process known as Technology Assessment. One component of such a process is Risk Assessment and involves the identification of the probability P_i of occurrence — over some relevant time interval — of a particular adverse consequence C_i associated with a specific device $D_i(t)$. The corresponding risk R_i is then formally defined as

$$R_i = P_i \times C_i, \tag{12.2}$$

and thus takes the form of a mathematical expectation of a particular event. In frequent applications, a specific adverse consequence is often implied and the risk R_i is then simply equated with the corresponding probability of occurrence P_i.

12.2.2 *Qualitative Evaluation*

While Eq. (12.2) is often useful for some assessment purposes, it suffers from two severe shortcomings:

(a) *Theoretical-Numerical Basis*
   It is not uncommon to have considerable disagreement on appropriate functions or numerical values for parameters to use in the calculation of the probability of occurrence P_i of some event of consequence C_i. Indeed, on some issues of emerging consumer interest, even expert estimates of P_i can vary by an order of magnitude and others may disagree on the choice of event consequences C_i.
(b) *Breadth of Considerations*
   On issues of intense societal interest, the relation (12.2) is often viewed as too restrictive. Members of a conceivably affected community look upon issues of opportunity and danger from a much broader perspective and include several factors associated with a device, some of which are suggested in Table 12.1.

The qualitative characterization of public concerns expressed in Table 12.1 also reflects on special apprehensions which many individuals may hold towards selective device developments. While physical and biological hazards and potential financial gains or losses are important, there may also exist deep anxieties about long-term consequent effects on life-style patterns, future restrictions, and community consequences. These effects on opportunities and danger are frequently interpreted as value issues or *value-laden*.

Table 12.1   Selected perception of factors which tend to increase or decrease societal apprehensions.

Factor	Increasing Societal Apprehensions	Decreasing Societal Apprehensions
Source of danger	Human error	Natural causes
Trust in advocates	Low	High
Exposure to hazard	Involuntary	Voluntary
Adversity prospects	Concentrated	Random
Fairness	Localized	Distributed

In order to circumvent many problems attributable to linguistic shadings, it is becoming increasingly important to distinguish between the concept of risk and the concept of safety. These terms are different and the following provides a convenient and effective definition:

(a) *Risk*

The term risk represents a *quantitative and objective expectation for a particular event* which may be computed by the product $R_i = P_i \times C_i$.

(b) *Safety*

The term safety is defined as a *qualitative and subjective personal judgment of the acceptance of a given risk.*

While engineers are assigned the responsibility of calculating the risk associated with a device $D(t)$, it is the collective judgment of a wider community which decides an issue of safety — and hence acceptance — of the device. Both of these considerations may be displayed as influence-seeking linkages in the engineering core progression,

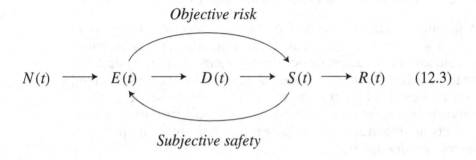

Objective risk

$$N(t) \longrightarrow E(t) \longrightarrow D(t) \longrightarrow S(t) \longrightarrow R(t) \qquad (12.3)$$

Subjective safety

illustrating also a basic cyclical feature which, conceivably, might never end unless some convergence evolves or is eventually terminated by some authority — or the underlying issues lose their significance.

Discussions of the extent of risk and safety associated with a particular prospective device are common concerns and may lead to heated debate and possibly even to a form of societal paralysis. Often overlooked in the assessment of opportunity and danger, is the risk and safety of not adopting a specific device in question; the retention of the status quo may present an entirely different set of opportunity and danger, and should also be considered.

12.3 Group Perspectives

Heterogeneous progressions associated with the development of some devices may involve a range of personal involvement, government regulations, financial resources, opportunity for community participation, and other specific interests. At one extreme, the device might be a rather benign development such as an improved garden tool or new type of handicapped mobility device. But at the other extreme, the device may be a hydroelectric power plant requiring extensive flooding of pristine valleys and relocation of villages, or it may be a costly and very risky resource extraction installation in an environmentally sensitive region. For large project undertakings, there may not be just one device $D_a(t)$ that is relevant but a conglomerate of variously developed devices $D_i(t)$ from which a selection is to be made. However, because of a variety of economical, political, and religious factors, a range of deeply held views may emerge to form a group characterization. Four broad classifications encompass most of the commonly held views.

12.3.1 *Technocratic*

The means of and capacity for the production and utilization of various devices is, in this perspective, taken to be a strength of society and its expanded development should be encouraged. An extreme form of this view holds that all political and economic government planning functions should be under the control of the technologically literate. This notion emerged in the early 1900s and sustained political parties which offered an explicit technocratic platform.

12.3.2 *Luddite*[†]

This label is currently taken to refer to the belief that particular devices may lead to dehumanizing and corrupting conditions and should therefore be abolished. It emerged in the early 1800s in England and involved workers who damaged factory looms as a protest against unemployment, skill obsolescence, inadequate working conditions, and low wages. Later, some Belgian workers would *accidentally* drop their wooden clogs, called sabots, into the delicate mechanism of spinning machines giving birth to the word *sabotage*.

12.3.3 *Apocalyptic*

The continuing production and uses of certain devices is, in this view, interpreted to imply an autonomous progression relentlessly leading to an out-of-control state of dire and calamitous consequences such as the collapse of social institutions, disintegration of communities, or destruction of the environment.

12.3.4 *Cautionary*

This view recognizes beneficial and non-beneficial aspects associated with device development and encourages a careful process of innovation, scale testing, prototype development, detailed assessment, and extensive consultation with experts, prospective users, and those directly affected.

These four societal perspectives on devices may be compactly summarized by the itemization

$$S(t) = S(t) \begin{Bmatrix} \textit{Technocratic} \\ \textit{Luddite} \\ \textit{Apocalyptic} \\ \textit{Cautionary} \end{Bmatrix}. \qquad (12.4)$$

Evidently further refinements to these categories may readily be identified; indeed, there exists a considerable literature on these topics, written from various perspectives.

[†]Conceivably named after Ned Ludd, England, a protest leader of the period 1811 to 1816.

12.4 Device Rejection

A common perception holds that devices are invented and adopted according to some loosely defined improve or enhance logic. History, however, suggest devices can be rejected or reversed as determined by a nominal societal or community preference. Here we list some illustrative examples.

Some early Tasmanian tribes had generally rejected bone tools and some Polynesian islanders chose not to make or use pottery. Kayaks had also, at one time, been dismissed by some polar Inuits. Beginning about 400 CE, Arab traders in North Africa replaced two-wheeled oxcarts by camels. Medieval society of western Europe introduced inadequately ventilated fireplaces, ignoring efficient heating ducts previously developed by the Romans.

In the early 1400s, China succeeded in building some very large ocean vessels, many ten times the size of Portuguese Caravels of the day. With a fleet of about 200 such ships, they explored the south Asian coastline and islands, sailing as far west as the Arabian Sea. Then, about 1450, their priorities changed and China halted all further oceanic explorations, even destroying many of their ship designs. Within decades, China had lost its ocean vessel construction and operating skills.

After Japan made contact with European mariners in the mid 1600s, the Japanese swordsmiths were quick to reproduce improved handguns. These were soon selectively used until the sword-wielding Samurai warrior class realized with much alarm that their honorable profession was under extreme threat of extinction as any unskilled peasant with a handgun could now be equally menacing. They petitioned their Government and soon handguns became too expensive or too difficult to acquire by the common citizenry. The Samurai subsequently recovered their supremacy and kept it alive for another 200 years. It was the sight of American gunships in Edo (now Tokyo) Bay in 1854 that convinced the Japanese to reinstate manufacture of handguns and other explosive weaponry.

Rejection of newer devices on secular grounds needs to be considered as distinct from those based on religious beliefs. For example, Islam declined to use the Gutenberg movable-type printing press for over 300 years, Judaism still requires that the Torah used in worship be handcopied by a ritual process, and many Old Order Amish continue to reject ownership — but not use — of modern appliances associated with electrical power.

Note that in a democratic society, any group can seek to influence the development or adoption of specific devices.

12.5   Market Penetration Trajectories

While the historical evolution of device invention parallels changing
insight on specialized aspects of matter and energy in meeting particular societal aspirations, market penetration of devices is invariably
influenced by a variety of other factors. These developments may also
occur on a variety of time scales and associated regional characterizations but, remarkably, there exist many cases which possess some invariant market penetration features representable by autonomous dynamical
formulations.

Consider as a first historical example some long-term device variations associated with water transport. The Sumerians and other early
riverine communities used poles to propel floating logs or rafts; crude
and eventually refined oars were the next stages of these evolutionary
devices. A major shift occurred when sails were used with the resultant
increases in both speed and carrying capacity, but requiring increasing
skill in judging wind speed and direction, and setting sails accordingly.
The Industrial Revolution provided for another substitutional change
namely that of burning coal on a large scale, thereby releasing the
energy residing in molecular bonding to heat water and generate steam
power; the resultant rotary propulsion thus provided freedom from the
vagaries of wind but this was attained at the price of requiring even
more human insight and technical skill — as well as releasing contaminating reaction products into the environment. The subsequent change
of replacing the steam engine and steam turbine by diesel engines provided some evident appeal of efficiency and convenience as demonstrated
by its general acceptance. Figure 12.2(a) provides a graphical indication of the evolution and extent of usage of motive power for water
transport.

A different characterization of land transport can be described based
on the total length of network for the transport of goods and people by
canal, railroad, and automotive road-transport, Fig. 12.2(b). Note that
within each of these three forms of ground transportation there exists
a large set of invention and innovations ranging from various types of
canal gates, to a variety of railway engines, and to an enormous diversity
of component devices for powered road vehicles.

A more specialized device dynamic can be associated with trans-
Atlantic passenger travel, Fig. 12.2(c). In the mid 1950s, piston driven
aircraft replaced steam ships and a decade later it was the jet aircraft which
added profoundly to total passenger transport. Concurrently, the aeronautical industry experienced an exceedingly large number of component

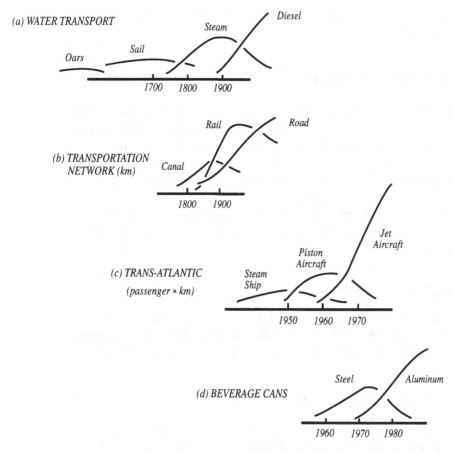

Fig. 12.2   Market entry trajectories for several generic types of devices.

device variations in its development — all invariably driven by a common interest in comfort, speed, cost, and convenience.

Finally a similar intrinsic dynamic is associated with common beverage cans. These devices became increasingly important some 30 years ago and, with little publicity, the steel used in manufacture was replaced by aluminum, Fig. 12.2(d).

Examination of the four cases illustrated in Fig. 12.2 makes evident a particular trajectory: new and improved devices replace older types and enter the societal usage market with what appears to be a characteristic S-shape trajectory. Such an S-shape market penetration has been observed to occur for many diverse devices in many regions and it is therefore worthwhile to explore an analytical-mathematical description of such a market competition/substitution feature for its potential use in device related forecasting.

12.6 Logistic Market Dynamics

The various differential and integrated factors associated with the appearance of new devices in a market place — labor requirements, corporate objectives, waste byproduct management, regulations, demands on scarce resources, public acceptance, environmental impact, etc. — are in the first instance specific to the type of device which enters a particular market place. Nevertheless, as indicated in Fig. 12.2, the general market penetration pattern commonly appears to possess an S-shape with the actual rate of entry and the magnitude of an eventual maximum highly case specific.

12.6.1 *Logistic Dynamics*

To begin, recall that the evolution of engineered devices targets the market place. This may be incorporated in the engineering context with $S(t)\{Market Place\}$ as an informing detail in the engineering core progression,

$$N(t) \to S(t) \to D(t) \to S(t)\{Market\ Place\} \to R(t). \qquad (12.5)$$

Then, in order to provide an intuitively clearer characterization, consider the market place as an aggregate cell which contains $X_m(t)$ number of a particular device $D(t)$ of interest at time t. These devices enter the market at a rate g_{+m} and depart at the rate g_{-m}. Hence, Eq. (12.5) can be expanded to read as shown in Fig. 12.3. The rate equation for $X_m(t)$ at any relevant time t is evidently given by

$$\frac{dX_m}{dt} = g_{+m}(t) - g_{-m}(t). \qquad (12.6)$$

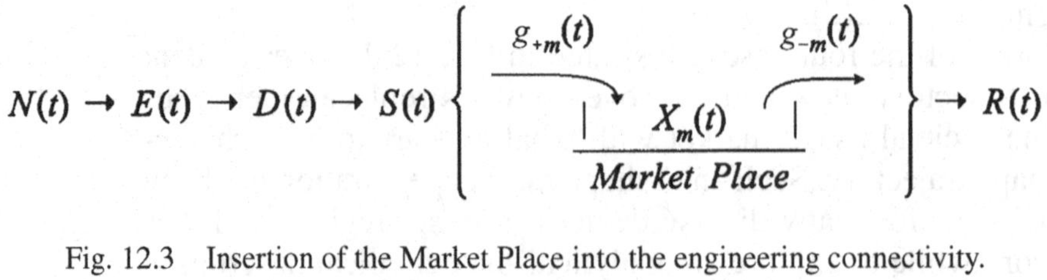

Fig. 12.3   Insertion of the Market Place into the engineering connectivity.

Consider next the specification of $g_{+m}(t)$ and $g_{-m}(t)$ in this equation. The exit rate $g_{-m}(t)$ may well be readily specified since with any device we may associate a particular average lifetime τ_m in the market place, subsequent to which it is considered a discard and destined for entry into the repository stockpile $R(t)$. For example, if $X_m(t)$ refers to automobiles then typically $\tau_m \simeq 12$ years, for television sets $\tau_m \simeq 6$ years, for commercial jets $\tau_m \simeq 25$ years, for ships $\tau_m \simeq 30$ years, etc.[†] Hence, to a general approximation, the exit rate may be represented by

$$g_{-m}(t) \simeq \frac{X_m(t)}{\tau_m}. \tag{12.7}$$

The entry rate g_{+m} is evidently more involved. Focusing upon the common interest for which an incipient number $X_m(t)$ already exists in the market place and for which evidence of a market interest is evident, we may write

$$g_{+m}(t) \sim A(\beta, X_m)X_m(t). \tag{12.8}$$

Here the functional $A(\beta, X_m)$ incorporates some autonomous but case specific effects relevant to the rate of device $X_m(t)$ marked entry. Three such heterogeneous engineering and societal causative factors are generally dominant:

(a) A proportional engineering-industrial-commercial capacity for the production and distribution of the device of interest.
(b) A proportional societal affinity to acquire these devices.
(c) A density-determined upper limit representing the eventual market saturation effect on the maximum number of $X_m(t)$ devices which the market can accommodate.

Under common manufacturing and acquisition conditions, a near equivalence is expected between the first two factors (a) and (b), and the saturation effect in (c) needs to involve $X_m(t)$ in subtractive form. These considerations therefore suggest the functional A of (12.8) as

$$A(\beta, X_m) \sim [\beta_1 + \beta_2 - \beta_3 X_m(t)], \tag{12.9}$$

[†] Variations in operational lifetime for specific classes of devices depend upon intensity of usage and maintenance schedules.

with each β_i a case specific positive parameter. Note the role of signs in affecting the rate dX_m/dt.

By substitution, the rate equation for $X_m(t)$, (12.6), now takes on the specific autonomous form of

$$\frac{dX_m}{dt} = [\beta_1 + \beta_2 - \beta_3 X_m(t)]X_m(t) - \frac{X_m(t)}{\tau_m},$$

$$= \left[\beta_1 + \beta_2 - \frac{1}{\tau_m}\right] X_m(t) - \beta_3 X_m^2(t)$$

$$= \alpha_1 X_m - \alpha_2 X_m^2, \tag{12.10}$$

with α_1 and α_2 as positive case specific parameters. This is a first order nonlinear differential equation which possesses two stationary points, i.e. $dX_m/dt = 0$ at $X_m = \{X_1^*, X_2^*\}$:

$$X_1^* = 0, \quad X_2^* = \alpha_1/\alpha_2. \tag{12.11}$$

Thus, the rate of market entry and its attainable maximum is highly case specific — the former depends separately on α_1 and α_2 and the latter on the ratio α_1/α_2. Figure 12.4 provides a graphical depiction of this equation with the domain of interest of $X_m(t)$ restricted to

$$X_m(0) = 0 < X_m(t) < \alpha_1/\alpha_2 = X_m(\infty). \tag{12.12}$$

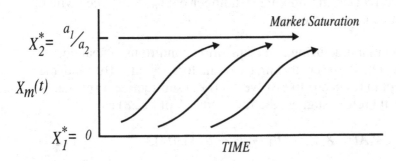

Fig. 12.4   Logistic market penetration $X_m(t)$ as a solution of Eq. (12.10b) for the domain of $0 < X_m(t) < \alpha_1/\alpha_2$. The cases displayed in Fig. 12.2 during times of increasing market penetration, i.e. $dX_m/dt > 0$, are arbitrarily displaced in time.

Note here the commonly observed S-shape market penetration trajectory; this component of the solution set of (12.10) is commonly called the Logistic.

A number of points need to be made about the above Logistic market penetration.

The Logistic function has been found to be a frequently useful market penetration representation. Cases for which this S-shape does not hold are those in which market penetration is subjected to effects not included as influencing factors of Eq. (12.10). Some such possible influences can readily be identified:

(a) Appearance of fatal operational flaws in the device (e.g. premature failure, . . .)
(b) Sporadic legislative action (e.g. unanticipated restrictions or taxation, . . .)
(c) Disruption of pertinent resource supply (e.g. war, labor strife, . . .)
(d) Substantial disruptions in manufacturing (e.g. product-line changes, corporate restructuring, . . .)
(e) Stimulated changes and societal preferences (e.g. emerging perception of hazards, changing societal attitude, . . .)
(f) Legal challenges (patent infringement, emerging liability, . . .)

Recall that such considerations are relevant to the use of the Scenario Planning and Delphi Technique of forecasting, Sec. 11.5.

Finally, the Logistic is commonly used during the early stages of market entry of a new device of interest when preliminary market data allows an empirical estimation of the parameter α_1 and α_2 in Eq. (12.10b). Projection can then be made including estimation of the maximum penetration $X_m(\infty)$, Fig. 12.4. Obviously, these estimates need to be updated as further market data becomes available.

12.6.2 Substitution Dynamics

As indicated, the Logistic applies to the device market entry phase — that is $dX_m/dt > 0$. What then can be said about the decline of $X_m(t)$ in the marketplace? Evidently when devices $X_m(t)$ depart from the market place, they are unlikely to leave a vacuum! That is, some substitution takes place and the replacing device enters the market according to a Logistic trajectory, Fig. 12.5, with its own unique set of parameters α_1 and α_2.

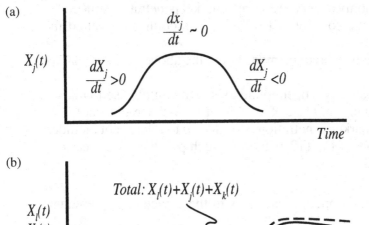

Fig. 12.5 Market penetration illustrating (a) single device entry-exit and (b) sequential device substitutions.

12.7 Population Dynamics

While in general a society may be characterized by a number of important attributes — literacy, technological optimism/pessimism, political perspective, consumer preference, gross domestic product, attitudes toward globalization, etc. — it is the total number of people in a market region which is of particular importance in affecting device related activities. Hence, an analysis of population projections — and some distinct characteristics — has long been of engineering interest.

12.7.1 *Population Dynamics*

Consider then $X(t)$ as the expected number of people in a given geographical region of interest. A change in this population ΔX during any time interval Δt is representable by a sum of distinct rate processes $(\Delta X / \Delta t)_i$ each of which add to or subtract from the population $X(t)$

during Δt:

$$\frac{\Delta X}{\Delta t} = \sum_i \left(\frac{\Delta X}{\Delta t}\right)_i. \qquad (12.13a)$$

These i-type processes will typically represent immigration (in-migration), emigration (out-migration), natality (birth), and mortality (death). With Δt arbitrary, the limit of taking this time interval progressively smaller yields the more convenient analytical form of a differential equation:

$$\frac{dX}{dt} = \left(\frac{dX}{dt}\right)_{immigration} - \left(\frac{dX}{dt}\right)_{emigration} + \left(\frac{dX}{dt}\right)_{natality} - \left(\frac{dX}{dt}\right)_{mortality}.$$

$$(12.13b)$$

A more compact notation gives

$$\frac{dX}{dt} = (g_+ - g_-) + (r_+ - r_-), \qquad (12.13c)$$

where g_+ and g_- are transboundary gain/loss flow rates while r_+ and r_- are birth and death rates. Note that the existence of transboundary flows defines an open system.

While the condition $dX/dt \neq 0$ of Eq. (12.13c) describes an instantaneous change in the population at some time t it also can represent a state of nonequilibrium. The converse $dX/dt = 0$ is by itself ambiguous because it might imply, for a period of time of interest, that $g_i = 0$ and $r_i = 0$ for all i or that their sums add to zero, i.e.

$$g_+ - g_- + r_+ - r_- = 0. \qquad (12.14)$$

Evidently, the former is of little relevance so that $dX/dt = 0$ based on some of the terms in Eq. (12.14) not equal to zero therefore defines the more relevant dynamic equilibrium of interest.

Immigration and emigration are processes associated with factors that are often difficult to project. These rates, however, may frequently be taken to be constant for specified periods of time or allowed to vary in order to characterize their effects on the evolution of $X(t)$. In contrast, both natality and mortality rates, r_+ and r_-, are generally expressed in

terms of observed fractional population changes per unit time according to

$$\left(\frac{\Delta X/X}{\Delta t}\right)_{natality} = \alpha_+ \qquad (12.15a)$$

to give, in the limit of $\Delta X/\Delta t \to dX/dt$, the birth rate as

$$r_+ = \left(\frac{dX}{dt}\right)_{natality} = \alpha_+ X. \qquad (12.15b)$$

By an analogous process for the population mortality rate we may take

$$r_- = \left(\frac{dX}{dt}\right)_{mortality} = \alpha_- X, \qquad (12.16)$$

so that Eq. (12.13c) may be written as

$$\frac{dX}{dt} \simeq (g_+ - g_-) + (\alpha_+ - \alpha_-)X$$
$$= g_0(t) + \alpha_0(t)X(t) \qquad (12.17)$$

as a sufficiently useful expression.

12.7.2 *Projections: Constant Coefficients*

The evident difficulty with projecting populations resides in the functions
$g_0(t)$ and $\alpha_0(t)$ of Eq. (12.17). For time intervals sufficiently short, these
terms may be taken as constant for which two cases are most informing:
Logistic population growth and the Malthusian Dilemma. We consider
the former first.

Human populations have in many developed countries and in recent
decades evolved under the influences of increasing standards of living,
expanded social equity, and broadened employment opportunity all tending to decrease the birth rate. While the aggregate effects on α_+ in
Eq. (12.17) are complex, this important feedback on $X(t)$ can be suggested by a weak natality-population relaxation given by

$$\alpha_+ \sim \alpha'_+ - \alpha'' X(t), \qquad (12.18)$$

where α'_+ and α'' are positive constants. Taking $g_0 = 0$ and substitut-
ing this intrinsic self-regulating feedback effect into Eq. (12.17) yields

an autonomous nonlinear dynamic given by

$$\frac{dX}{dt} \sim (\alpha'_+ - \alpha''_+ X - \alpha_-)X$$

$$\sim (\alpha'_+ - \alpha_-)X - \alpha'' X^2. \tag{12.19}$$

Note its functional equivalence to the Logistic if $\alpha'_+ > \alpha_-$, (12.10).

In analyzing populations which obey a Logistic dynamic, the equilibrium population of (12.19)

$$(\alpha'_+ - \alpha_-)X - \alpha'' X^2 = 0 \tag{12.20a}$$

gives two stationary points:

$$X_1^* = 0, \qquad X_2^* = \left(\frac{\alpha'_+ - \alpha_-}{\alpha''}\right). \tag{12.20b}$$

With $\alpha'' > 0$, the parameter space is now determined by whether $\alpha'_+ > \alpha_-$ or $\alpha'_+ < \alpha_-$. Using again a graphical interpretation of Eq. (12.19), a solution flow in $\alpha'_+ - \alpha_-$ Cartesian parameter space results and is depicted in Fig. 12.6. Evidently, the space defined by $\alpha'_+ > \alpha_-$ yields a stable and physically meaningful attractor regardless of whether

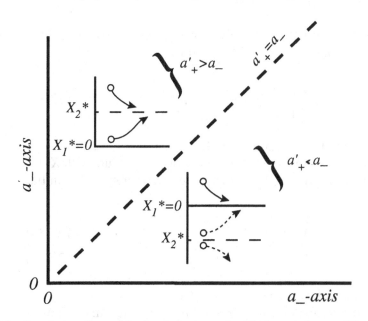

Fig. 12.6 Population flow of the Logistic dynamic $dX/dt = (\alpha'_+ - \alpha_-)X - \alpha''_+ X^2$ displayed on an α'_+ and α_- Cartesian plane.

the initial condition is above or below this stationary point X_2^*. In the remaining parameter space $\alpha_+' < \alpha_-$, the population is depletive. We again discard the physically meaningless range of negative population numbers.

About 200 years ago, Thomas Malthus, England, publicized some views concerning his expectations of the continuing growth of the human population in light of the simultaneous availability of the food supply: the former seemed to increase exponentially with time while the latter, being dependent upon the increase in arable land, tended to increase linearly with time. Evidently, if these trends continued, famine and widespread social disorder would eventually occur. These views have been influential ever since.

If $X_1(t)$ is taken to represent the population and $X_2(t)$ the supply of food — each in appropriate units — then the relevant simultaneous dynamical description is given by two equations with constant coefficients:

$$\frac{dX_1}{dt} = \beta_1 X_1, \quad \beta_1 > 0, \tag{12.21a}$$

$$\frac{dX_2}{dt} = \beta_2, \quad \beta_2 > 0. \tag{12.21b}$$

Further, if some minimum food supply per person was required then these dynamical descriptions are valid only for $t \leq t_x$ for which

$$\frac{X_2(t)}{X_1(t)} \geq \left(\frac{X_2}{X_1}\right)_{crit}. \tag{12.22}$$

This Malthusian Dilemma is graphically described in Fig. 12.7.

Of some interest is the time t_x for this catastrophe to occur. With $X_1(t)$ and $X_2(t)$ uncoupled in Eq. (12.21), integration gives for specified initial conditions

$$X_1(t) = X_{1,0}\, e^{\beta_1 t}, \tag{12.23a}$$

$$X_2(t) = X_{2,0} + \beta_2 t, \tag{12.23b}$$

to yield a transcendental expression for $t = t_x$.

Malthus' dire predictions have subsequently generated much apprehensive interest since, for many countries, the annual fractional increases in human population actually exceeded the annual fractional increases of

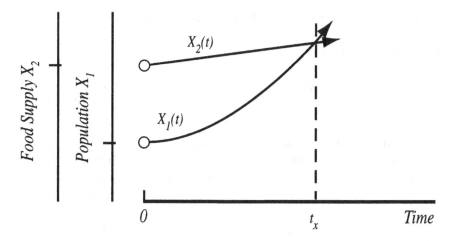

Fig. 12.7   Graphical depiction of the Malthusian Dilemma.

arable land. Indeed, improved health care beginning in the mid 1900s significantly reduced infant mortality and increased longevity, contributing thereby to a doubling of the global population in only $\sim$40 years, 1950 to 1990. But, to compound Malthus' hypotheses, the need for additional arable land vanished after $\sim$1950 because of the following agricultural and engineering developments:

(a) Manufacture of low-cost synthetic fertilizer
(b) Introduction of improved yield and more disease-resistant crops
(c) Expanded practice of efficient irrigation

Finally, by the 1990s, decreasing natality — especially in regions with an historically high fertility — became evident with the consequence of much reduced global population growth rate. It is now commonly expected that a global population near-equilibrium, i.e. $r_+ \simeq r_-$, at $\sim$10 billion people might come about by 2050.

12.7.3   *Projections: Time-Dependent Coefficient*

For longer regional projections, it is necessary to retain a time dependence on the various parameters so that Eq. (12.17) may be written as

$$\frac{dX}{dt} = g_+(t) - g_-(t) + [\alpha_+(t) - \alpha_-(t)]X(t)$$

$$= g_0(t) + [\alpha_+(t) - \alpha_-(t)]X(t). \qquad (12.24)$$

Of interest now are two cases involving time-dependent coefficient functions.

For the elementary dynamic

$$\frac{dX}{dt} = g_0(t) \tag{12.25}$$

and supposing that during some time interval of interest, the net migration f_0 changes monotonically from some negative value $f_0 < 0$ through $f_0 = 0$ to some positive $f_0 > 0$, this transformation may be represented by

$$g_0(t): - \rightarrow +. \tag{12.26}$$

Evidently, the population $X(t)$ will begin at some $g_0 < 0$ with g_0 increasing until $g_0 > 0$ and the corresponding vector — that is dX/dt of Eq. (12.24) — will similarly increase, pass through zero, and eventually become positive. This transition and its inverse are illustrated in Fig. 12.8.

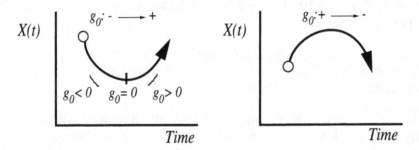

Fig. 12.8   Population flow of the dynamic $dX/dt = g_0(t)$ for the monotonic transformations $g_0(t): - \rightarrow +$ and $g_0(t): + \rightarrow -$ as a function of time.

An example in which an individual parameter may vary monotonically with time, though the effect need not necessarily be monotonic, is provided by the case of natality and mortality parameters varying independently. For this case, the nonautonomous extension may be written as

$$\frac{dX}{dt} = [\alpha_+(t) - \alpha_-(t)]X$$
$$= \alpha_0(t)X. \tag{12.27}$$

A phenomenon observed in countries undergoing rapid industrialization is that mortality rates decrease initially much more rapidly than natality. Interestingly, the reason for a relatively more rapid decrease in $\alpha_-(t)$ is based on devices, especially the availability of improved medical facilities, clean water supply, and comprehensive sewage installations, leading therefore to a decrease in $\alpha_-(t)$; in contrast, decreases in $\alpha_+(t)$ tend to be determined by the delayed effects of cultural adjustments and overall economic conditions. The functions $\alpha_+(t)$ and $\alpha_-(t)$, and hence $\alpha_0(t)$, tend to vary with times as suggested in Fig. 12.9(a), so that in parameter space $\alpha_0(t)$, the transformation of $\alpha_0(t)$: $+ \rightarrow -$ yields a

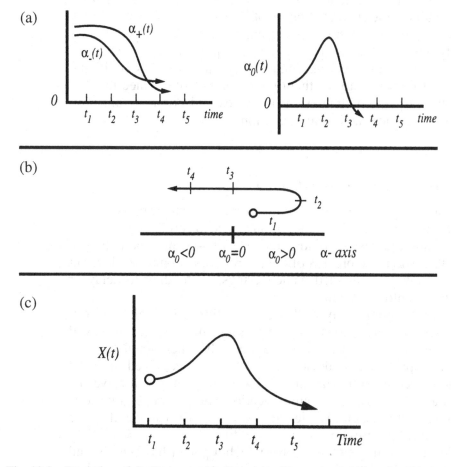

Fig. 12.9   Depiction of the Demographic Transition illustrating $\alpha_+(t)$ and $\alpha_-(t)$ time dependencies, monotonic transition of α_0: $+ \rightarrow -$, and the consequent population evolution.

nonmonotonic variation, Fig. 12.9(b). The corresponding solution for this population dynamic is suggested in Fig. 12.9(c) and often labeled the Demographic Transition.

According to this demographic perspective, industrialization and commercialization in a less developed country leads to a worsening of population density before it gets better, Fig. 12.9(c).

12.8 Commercialization and Wealth

We had previously noted that economics and politics are two of three dominant factors in the characterization of society $S(t)$, Sec. 12.1. These two factors appear increasingly important for they relate to commercial activity and wealth. The progression component $\cdots \rightarrow D(t) \rightarrow S(t) \rightarrow \cdots$ of the engineering connectivity thus contains additional information of relevance to the theory and practice of engineering. We recognize two aspects of interest: (a) devices and commercialization intrigue and (b) devices and wealth creation.

12.8.1 *Devices and Commercialization*

It is a frustrating experience for inventors to discover that the process from invention to significant market penetration can be torturous — especially if it appears that the device could become a high-profit generator. The reason for this is that the process of commercialization may well attract hijackers and involve extended maneuverings, all tending to delay or diminish the inventors' benefits.

Significant inventions may well occur at a time when relevant government regulatory provisions may be lacking; indeed, one may well ask how regulations could be in place since the operational features and societal impact of new devices are not fully recognized until the latter stages of commercialization. Well known examples involve the telegraph, radio, laser, satellite communication, computer operating software, genetic materials, etc. While each case is distinct in detail, some features of typical progressions may be identified.

Consider now some new device for which a patent has been issued and the process of final development and initial stages of commercialization have taken place. By then some hijackers have become alerted and proceed with a parallel competing progression involving some device variations:

$$\begin{array}{ccccc}
\textit{Initial} & & & \textit{Primary} & \textit{Stages of primary} \\
\textit{progression} & \!\!\!\!:\ldots\ldots\textit{Patent} \rightarrow & \textit{development} & \!\!\!\!\rightarrow & \textit{commercialization} \\[2mm]
\textit{Competing} & & & \textit{Emerging} & \textit{Competing} \\
\textit{progression} & \!\!\!\!:\cdots\cdots\cdots\cdots\cdots & \textit{hijackers} & \!\!\!\!\rightarrow & \textit{commercialization}
\end{array}$$

(12.28a)

In the absence of clear government legislations, a legal and/or public relations process is initiated by the primary developer and directed to the hijackers; note, however, that the hijackers may well claim that — in the absence of specific and direct laws — they are not doing anything wrong by their reasonable interpretation of existing laws. While legal and other maneuverings proceed, the resultant competition may well lead to both public confusion and market uncertainty:

$$\textit{Patent} \rightarrow \begin{array}{c}\textit{Primary} \\ \textit{development}\end{array} \rightarrow \textit{Commerce I} \searrow$$
$$\begin{array}{c}\textit{Market} \\ \textit{uncertainty}\end{array}$$
$$\begin{array}{c}\textit{Emerging} \\ \textit{hijackers}\end{array} \rightarrow \textit{Commerce II} \nearrow$$

(12.28b)

Eventually new legislation is introduced or existing laws revised — or a clarification is issued by the courts — and market order established with all agents actively pursuing profit and an expectation of increased societal prosperity:

$$\textit{Patent} \rightarrow \textit{Primary} \rightarrow \textit{Commerce I} \searrow \qquad \begin{array}{c}\textit{Gov't legislation}\end{array} \searrow$$
$$\textit{Chaos} \rightarrow \begin{array}{cc}\textit{Market} & \textit{Societal} \\ \textit{order} \rightarrow & \textit{prosperity.}\end{array}$$
$$\textit{Hijacker} \rightarrow \textit{Commerce II} \nearrow \qquad \begin{array}{c}\textit{Judicial} \\ \textit{intervention}\end{array} \nearrow$$

(12.28c)

In general such a progression may be very uncertain in detail, all depending upon specifics of device features, public impact, legal issues, and market dynamics. In practice, the associated market competitiveness has often been a considerable stimulant to both device innovation and economic activity.

12.8.2 *Devices and Wealth*

Engineers invariably incorporate commercialization of devices in their
activities because of a particular and universal societal interest: the production and sustainment of wealth. Indeed, there exists a subtle and significant connection between wealth and the engineer's capacity for device
invention and innovation. But first, what is wealth?

Classical economic theory asserts that wealth occurs as a result of
economic growth, and this growth requires continuing division of labor
in manufacturing and expanded capital investment in productive industrial facilities. In terms of a heterogeneous progression, this may be suggested by

$$\left\{ \begin{array}{c} \textit{Division of labor and} \\ \textit{capital investment} \end{array} \right\} \rightarrow \left\{ \begin{array}{c} \textit{Economic} \\ \textit{growth} \end{array} \right\} \rightarrow \{\textit{Wealth}\}. \qquad (12.29)$$

This classical view of wealth creation has in recent years been found
deficient for two reasons:

(a) *Diminishing Returns*
    Continuing division of labor and capital investments is known
    to eventually yield diminishing economic returns. Since many
    economies have continued to sustain substantial growth over long
    periods of time it is concluded that there must be other factors of
    importance. Indeed, research suggests that increases in the division
    of labor and investments account for perhaps only ~25% of long
    term economic growth.
(b) *Invention and Innovation*
    Classical economic theory provides little recognition of the role of
    invention and innovation associated with devices. Recall that our
    definition of a device includes both hardware/object as well as soft-
    ware/cognition entities.

If invention and innovation contribute to the remaining ~75% of
wealth creation, what then is it about these inventive and innovative
devices and entities that render them of such great economic importance,
especially in contemporary times? Four factors appear to be particularly
dominant:

(a) *Added Value*
    Continuing device improvements — by invention or innovation —
    may make existing devices more valuable. For example, catalytic
    converters add value to a conventional automobile, and diesel-electric

locomotives are more effective to a railroad business than steam locomotives.

(b) *Increased Labor Productivity*

Ingenious ideas and device specialization have added considerably to labor productivity in manufacturing and in the service sector. Examples are specific robots in mass production and advanced sensors for polymer production.

(c) *Repeated Usage*

Once invented, a device may find repeated adaptations. While microwave generators first proved very important in radar and telecommunications, such devices have since proven to be useful in subsequent microwave ovens; similarly, microprocessors designed for computers are now widely used in numerous industrial, transportation, and domestic process-control systems.

(d) *First-Use Stimulation*

The first-use of an innovative concept or device can have a profound effect on subsequent activities of a corporation[†]. One of the more recent examples of such first-use stimulation is associated with integrated circuits and computer operating systems.

Evidently, a contemporary characterization of wealth creation is now more accurately suggested by a heterogeneous progression involving specifically the inventive and innovative activities of engineers. Noting that innovation includes many elements of commercialization — economics of scale, novel marketing, specialized adaptations, etc. — relation (12.29) may hence be more appropriately written as

$$\left\{\begin{array}{c} division\ of\ labor, \\ capital\ investment, \\ invention, \\ and\ innovation \end{array}\right\} \rightarrow \left\{\begin{array}{c} sustained\ economic \\ growth \end{array}\right\} \rightarrow \{wealth\}.$$

$$(12.30)$$

These various considerations have in recent years spawned much analysis introducing also concepts such as *technological change, wealth-creating strategies,* and *creative economies.* And evidently, engineers are important participants in such developments.

[†]The assumption here is that there exist no unrecognized fatal device flaws or disastrous downstream effects (e.g. deHavilland Comet, hydrogen airships, DDT insecticide, asbestos insulation, etc.).

12.9   Peculiar Imbalance

The engineering connectivity context places society $S(t)$ between newly produced devices $D(t)$ and the spent or obsolete device repository $R(t)$, i.e. $\cdots \rightarrow D(t) \rightarrow S(t) \rightarrow R(t)$. That is, society is the acceptor of new devices and the disposer of used devices.

In general, participants of societal institutions such as the military, commerce, hospitals, etc., and most individuals, display a much greater interest in the acquisition of new devices than in the unavoidable destination of obsolete devices. Outdated refrigerators, worn tires, old aircraft, automoded combat vehicles, plastic shopping bags, inefficient factory equipment, dated medical diagnostic facilities, older-style PCs, etc., are expected to *just disappear* with little adverse consequence. That is, while the acquisition linkage $D(t) \rightarrow S(t)$ can greatly fascinate individuals and societal institutions, the companion disposal linkage $S(t) \rightarrow R(t)$ is commonly of little interest or even an annoyance.

Recent publicity about saturated landfill sites, toxic-pond leakage, and visible junk piles, have contributed to a greater awareness not only of the existence and saturation of the residual inventory $R(t)$ but also to the $S(t) \rightarrow R(t)$ disposal process, and hence to the initial propensity for new devices $D(t) \rightarrow S(t)$. We consider some related analyses about this *end-of-pipe* subject — also sometimes called the *back-end* of the material flow cycle — in the next chapter.

12.10   To Think About

- Society often attributes device successes to some individual or group of individuals, and failures are often blamed on *the system*. Critically examine such statements.
- Itemize some historic and recent technical blunders which occurred because of incorrect assumptions about societal attitudes.
- Equations commonly derived for physical problems, say thermo-dynamics, electronics, fluid flow, etc., imply large numbers of atoms, electrons, and molecules. Might one therefore think of Eq. (12.10) as having similar validity if the implied number of producers and consumers in the market place of interest is sufficiently large?

CHAPTER 13

Repository: Inventory and Projections

$$\cdots \to S(t) \to R(t)$$

13.1 Material Metabolism

The Contemporary Engineering core connectivity $N(t) \to E(t) \to D(t) \to S(t) \to R(t)$ incorporates a flow of materials and energy. Indeed, with the critical intellectual characterization of $E(t)$ and $S(t)$, we may assert the creation and transmission of knowledge and ingenuity also becomes an intrinsic component of the core connectivity. Focusing for now on material flow, we note that all materials have their beginning in nature $N(t)$, then pass through various transformations, with spent devices and recoverable materials eventually deposited in some repository $R(t)$. The various material flow components are associated with the following:

(a) Material resource extraction and primary smelting: $N(t) \overset{\uparrow}{\to} E(t)$
(b) Secondary processing, component manufacture, and device assembly: $E(t) \overset{\uparrow}{\to} D(t)$
(c) Device distribution and use: $D(t) \to \overset{\uparrow}{S}(t)$
(d) Eventual device disposal: $S(t) \to R(t)$

These classes of material transformations involve several variable but ultimately reducible waste material flow components:

(a) Non-recoverable wastes leading to the polluting of air, water, and soil: $\overset{\downarrow}{N}(t)$
(b) Recoverable sequestered material waste: $\overset{\downarrow}{R}(t)$

(c) Discards as depositions in some repository: $\rightarrow R(t)$
(d) Partial repository content becoming identified with contamination of nature: $N(t)$
$\uparrow$

Figure 13.1 provides a graphical depiction of this material flow as a specialization of the engineering connectivity appropriate to Contemporary times.

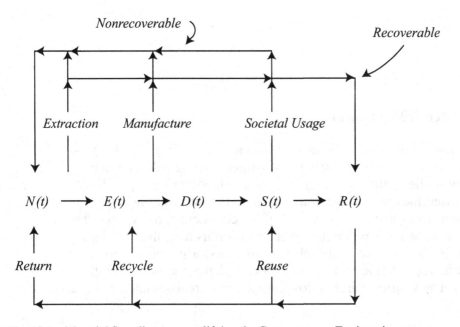

Fig. 13.1   Material flow diagram amplifying the Contemporary Engineering core connectivity $N(t) \rightarrow E(t) \rightarrow D(t) \rightarrow S(t) \rightarrow R(t)$.

13.2   Repository Inflow

As suggested in Fig. 13.1, three distinct component processes contributing recoverable and non-recoverable wastes to the repository $R(t)$ are identified: extraction processes, manufacturing processes, and societal-usage processes. Note that there exist numerous evolving regional and national regulations on the quantities and rates of production of certain process byproducts, imposing thereby a variety of manufacturing restrictions.

13.2.1 *Extraction*

Materials for all devices have their beginning in the resource extraction industry consisting of strip mining, open pit mining, subsurface mining, forestry, and the textile and leather industry. As noted in Sec. 10.4, the natural low enrichment of many desirable elements and minerals demands that large quantities of soil and rock be removed, transported, and selectively processed; as a rule of thumb, each one mass unit of eventually useful mined matter for a device requires — on average — more than 100 mass units of terrestrial materials to be moved, mechanically processed, and chemically treated; thus, substantial accumulations of tails, slag, dust, chips, and assorted sludge of varying toxicity accumulate at various extraction and processing sites.

Additionally, some materials such as residual sulfides in tails and slag, may combine with atmospheric oxygen and processes of erosion, leading thereupon to acidic runoff and leaching. Evaporation and leakage from toxic holding ponds also contributes to hazardous material migration. Finally, the use of heavy machinery and transportation vehicles unavoidably contributes to various solid and liquid wastes, as well as gases such as CH_4, CO_2, NO_x, and SO_2; all are released into the environment and largely considered non-recoverable waste streams.

13.2.2 *Manufacturing*

Useful materials from extraction and smelting provide raw materials for manufacturing. The manufacturing industry encompasses myriad distinct material transformations — chemical treatment, metallurgical processing, cutting, stamping, shaping, welding, casting, milling, molding, and polishing — and finally device assembly.

At each of the many manufacturing stages, further material wastage occurs and energy degradation takes place. Some of these losses consist of non-recoverable gases and liquids which enter the atmosphere and the earth's surface. Other losses, such as some liquids and most solids, become sequestered material wastes contributing to the repository inventory $R(t)$ of scrap heaps, settling ponds, holding tanks, and junk piles; waste gases are either discharged into the atmosphere or further processed into compound liquids or solids for sequestering.

Of particular safety interest in manufacturing are the legislatively regulated materials. This includes not only the well known biological hazardous substances such as DDT and radioactive materials such as spent nuclear reactor fuels, but other important and restrictive materials such as CFCs (chlorofluorocarbons), PCBs (polychlorinated biphenils), CCAs (copper, chrome, arsenic), dioxins (heterocyclic hydrocarbons), and compounds of heavy elements such as lead, mercury, and cadmium.

13.2.3   *Societal Usage*

The purpose of the material flow and transformations depicted in Fig. 13.1 is to provide devices for a societal interest. This interest consists not only of utility ensuing from personal usage but also due to numerous sectorial interests: business and commercial equipment, military weaponry, administrative aids, industrial machinery, tools for the service sector, etc. That is, devices may be taken to represent a large set of objects used by individuals and communities because they are endowed with a measure of utility.

Uses of these devices invariably requires energy, leads to wear, and contributes to component burnup — in short, it leads to degradation of energy and material assemblies. As indicated in Fig. 13.1, both non-recoverable process wastes occur as do sequestered material wastes and device discards; the former leads to general environmental contamination while the latter initially appear as contributors to scrap heaps, garbage piles, material separation facilities, and other sequestering installations all of which are components of the distributed repository $R(t)$.

13.2.4   *Numerical Estimations*

Increasing effort has in recent years been extended to seek estimates of the various material flow rates. For the flows suggested in Fig. 13.1, the following appear to be mid-range global quantities:

(a) *Primary Supply*
    The global total extraction of elements, minerals, and hydro-carbon fuels (coal, petroleum, natural gas) which enters the process stream amounts to $\sim$20 Tera[†] kg/year; this is equivalent to a global per capita of $\sim$10 kg/day.
(b) *User Discards*
    Device discards amount to about 10 Tera kg/year as direct inflow to the worlds slag piles, toxic ponds, scrap heaps, junk piles, landfill sites and garbage piles; that is $\sim$5 kg/day per capita.

[†]1 Tera $= 10^{12}$.

(c) *Non-recoverable Waste*

The global nonrecoverable process wastes are about 5 Tera kg/year approximately equally divided between gases — mostly CO_2 from hydrocarbon combustion — and liquids and solids; this translates to a per capita global impact of $\sim$2.5 kg/day — all of which becomes part of degraded nature $N(t)$.

(d) *Sequestered Waste*

Various estimates suggest the sequestered global wastes to be $\sim$5 Tera kg/year contributing $\sim$2.5 kg/day per capita to the repository $R(t)$.

These quantities are suggested in graphical form in Fig. 13.2. Note that on average, about 7.5 kg per person accumulates in the world's garbage and waste piles each day. It is, however, a good rule to keep in mind that global per capita values of any quantity — economic, caloric, life expectancy, etc. — mask substantial regional and material differences; for example discard accumulations for Western Europeans are about three times the global average while for North Americans they are about five times the global average.

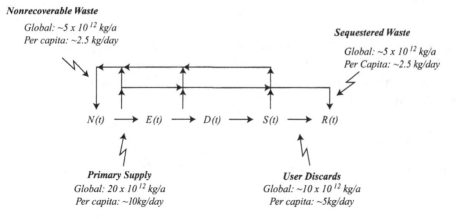

Fig. 13.2 Material flows associated with the engineering connectivity progression.

13.3 Repository Outflow

In a finite world, no repository $R(t)$ can sustain an indefinite matter inflow — a controlled outflow needs to be introduced or overloading consequence are inevitable. Three classes of societally influenced outflow streams from the repository $R(t)$ are identified, Fig. 13.1.

13.3.1   *Reuse Destinations:* $S(t) \longrightarrow R(t)$

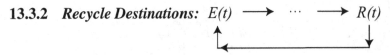

Device reuse can be classified by three principal categories:

(a) *As-Is*
   Examples are the well established markets for used automobiles, used furniture, used clothing, used machinery, etc.
(b) *To-Be-Altered*
   There exist many devices which are altered and then used for related purpose: passenger aircraft for freight transport, obsolete PCs useful for alternate reduced-function purposes, inefficient factories converted to warehouses, suitably located warehouses as *avant-garde* housing, etc.
(c) *Preservation*
   A small specialized reuse market exists for purpose of public display, historical preservation, as well as private heritage collections.

It is estimated that about 10% of spent devices are reused at least once. Note that this reuse activity has the effect of increasing the device mean residence time in the market place, but ultimately they still contribute to the repository inventory $R(t)$.

13.3.2   *Recycle Destinations:* $E(t) \longrightarrow \cdots \longrightarrow R(t)$

Device disassembly and material separation can provide for manufacturing raw materials which are indistinguishable from — or near-equivalent to — those provided by the resource extraction industry; additionally, components of some spent devices can be used as replacement parts of others. This applies especially to those materials and devices which bear a high cost of primary extraction or otherwise are rare. There also exist increasing prospects for recycling discarded devices especially if material distribution or compaction is risky or hazardous.

A total of about 25% of all sequestered wastes and discards are typically destined for recycling.

13.3.3 *Return Destinations:*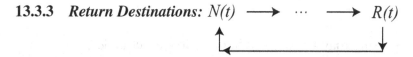

Common yard wastes such as grass cuttings, tree branches, and decaying fibrous materials may well be returned to nature — preferably mixed with clean rock and concrete — as most suitable landfill. Future residential and industrial land may thus be generated.

13.3.4 *Retain Destinations:* $\overset{\frown}{R(t)}$

Troublesome waste accumulations are those which decay or decompose very slowly requiring simultaneously or variously chemical/heat/pressure treatment. Among these are paints, treated lumber, plastic wrappings, styrofoam, and special thermal/electrical insulators. Some of these materials may be retained in domains which seem environmentally inert in the intermediate term and — with time — further reproceed toward a less hazardous state.

In a separate category are those accumulations which require sequestering on time scales of centuries or even millennia. These are wastes such as highly radioactive materials and potent biological toxins. Very special and costly treatments involving nuclear transmutation and molecular reconstruction may be introduced to reduce the volume or toxicity of such materials.

This retain category suggests the need for considerable device development which contribute to the transformation of hazardous materials into contained or harmless form. The alternative is to develop sequestering devices of great reliability and substantial longevity.

13.3.5 *Restoration Programs:* $\overset{\frown}{N(t)}$

Increasing emphasis has in recent years been placed on the restoration of despoiled landscapes. Typical cases include

(a) Depleted surface mining sites
(b) Obsolete industrial processing sites
(c) Former open toxic-storage sites
(d) Inactive logging and other roads

The common objective then centers on returning these lands to *greenfield* conditions. Several points need to be noted in this process of land restoration:

(a) Novel earth processing may be required in order to extract undesir-
    able impurity components
(b) Land surface restoration may be selectively and differentially under-
    taken in the interests of biodiversity
(c) Careful restoration and land management may yield broadly based
    hydrological benefits such as ground water replenishment, flood con-
    trol, and aquatic wetland provisions

Invariably a more integrated environmental perspective is required for a
successful restoration.

13.3.6   *The Multiple-R Maxim*

Popular usage has assigned a managerial maxim to contemporary material flows: reduce, reuse, recycle, retain, return, and restore. These terms
may be associated with matter-flow components of the core engineering
connectivity:

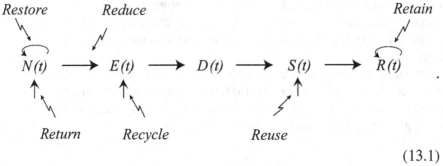

$$\tag{13.1}$$

Continuing ingenuity is thus expected of engineers.

13.4   Repository Stockpile Projections

The eventual appearance of device discards in the repository $R(t)$,
together with recoverable wastes from extraction, manufacturing, and
societal usage processes involved in producing and using devices, is
implicit in the evolution of the heterogeneous engineering progression,
Fig. 13.1. To this depictions it is useful to add an insert which is to
suggest an integrated repository cell into which all device discards and

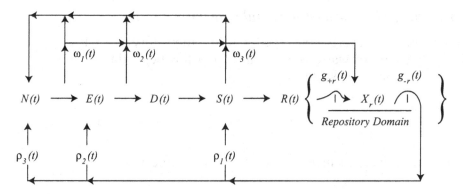

Fig. 13.3   Insertion of repository detail into the engineering connectivity.

recoverable process wastes enter and from which a partial or full exit eventually needs to be provided; the variable $X_r(t)$ is taken to be the quantity of waste in this repository $R(t)$ at any time t, Fig. 13.3. Here, $g_{+r}(t)$ is the total entry rate of wastes into the repository and $g_{-r}(t)$ is its exit rate. The entry rate $g_{+r}(t)$ consists of the device exit rate from the market $g_{-m}(t)$, (12.7), together with a fraction γ_i of the waste process rates $\omega_1(t)$, $\omega_2(t)$ and $\omega_3(t)$ which are retained in the terrestrial remains storage facilities. Then, the device exit rate $g_{-r}(t)$, is equivalent to the sum of rates associated with the processes of reuse $\rho_1(t)$, recycle $\rho_2(t)$, and return $\rho_3(t)$. That is

$$g_{+r}(t) = \frac{X_m(t)}{\tau_m} + \sum_i \gamma_i \omega_i(t), \tag{13.2a}$$

and

$$g_{-r}(t) = \rho_1(t) + \rho_2(t) + \rho_3(t), \tag{13.2b}$$

so that the dynamical equation which governs the time evolution of $X_r(t)$ in the repository is written in general form as

$$\frac{dX_r}{dt} = g_{+r}(t) - g_{-r}(t)$$
$$= \frac{X_m(t)}{\tau_m} + \sum_i \gamma_i \omega_i(t) - \rho_1(t) - \rho_2(t) - \rho_3(t). \tag{13.3}$$

Of evident interest is the time variation of $X_r(t)$ in this integrated repository $R(t)$ for three sequential time intervals beginning with $t = 0$, representing the time of first appearance of a repository deposition.

13.4.1 *Beginning of Repository Accumulations:* $0 \leq t < t_1$

It is common not to be overly concerned with the repository exit rates during the early stages of waste accumulation. For this beginning time interval $0 \leq t < t_1$ one may write

$$\frac{dX_r}{dt} \simeq g_{+r}, \qquad (13.4)$$

with a likely gradual increase in this entry rate g_{+r} with time, Fig. 13.4(a).

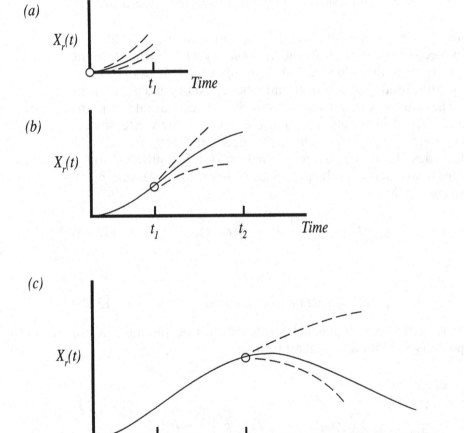

Fig. 13.4 Evolution of repository $R(t)$ accumulation for three characteristic time intervals. (a) beginning, (b) midlife, and (c) resolution.

13.4.2 *Midlife Accumulations:* $t_1 < t < t_2$

With an increasing accumulation of wastes in the repository $R(t)$, a correspondingly increasing societal awareness and apprehension results for a number of reasons:

(a) *Aesthetics*
Increasing recognition of the unsightliness of waste accumulations.
(b) *Health*
Health hazards may be identified with different forms of wastes: hazardous liquids may flow or seep into the ground and poison aquifer water supplies, poisonous gases may escape into air thereby contributing to respiratory risks, some dilapidated structures may pose fire hazards and also attracts rodents, and moisture accumulations may provide breeding grounds for disease-carrying mosquitos.
(c) *Resources*
Scarce materials might be contained within the wastes such as rare-earth metals and special alloys, all possessing commercial value.

Each of these apprehensions can be addressed by a reduction of the quantity of repository accumulation $X_r(t)$. The three types of exit rates are summarized as follows:

(a) *Reuse:* $\rho_1(t)$
Many devices in $R(t)$ may be reused in specialized parts of the market place (e.g. good used cars, good used clothing and furniture, older but still certifiable commercial aircraft, modified devices for other uses, collectibles, ...).
(b) *Recycle:* $\rho_2(t)$
Many spent devices may be in a form suitable for ready recycle (e.g. glass containers, aluminum beverage cans, newsprint, etc.) while others may require considerable material stream separation (e.g. wrecked cars, electrical appliances, building materials, heavy equipment, etc.). We add that some devices are now specifically designed for ease of disassembly to aid recycle.
(c) *Return:* $\rho_3(t)$
Some devices and process stream wastes may contain components readily releasable into the environment; this invariably involves biodegradables (e.g. untreated wood, non-dyed textiles, etc.) and natural minerals and elements.

For this midlife period of interest it is now necessary to include an exit rate $g_{-r}(t)$ from the repository $R(t)$, composed of the sum of the individual exit rates $\rho_1(t)$, $\rho_2(t)$, and $\rho_3(t)$. Taking these rates to consist of some collective percentage decline per unit time for this $t_1 < t < t_2$ interval suggests

$$\frac{dX_r}{dt} = \frac{X_m(t)}{\tau_m} + \sum_i \gamma_i \omega_i(t) - \sum_j \rho_j(t)$$

$$= g_{+r}(t) - \sum_j \delta_j(t) X_r(t)$$

$$= g_{+r}(t) - \delta_r(t) X_r(t), \tag{13.5}$$

with $g_{+r}(t)$ and $\delta_i(t)$ slowly increasing with time. As indicated in Fig. 13.4(b) and depending upon the magnitude of the net exit rate parameter $\delta_r(t)$, various decreases in dX_r/dt result but they all project $X_r(t)$ towards an asymptote exceeding $X_r(t_1)$; from (13.5), this asymptote is evidently

$$X_r(\infty) \sim \frac{g_{+r}(\infty)}{\delta_r(\infty)}. \tag{13.6}$$

Reducing $g_{+r}(\infty)$ or increasing $\delta_r(\infty)$ will therefore reduce the magnitude of the asymptote $X_r(\infty)$.

13.4.3 *Attempts at Resolution: $t > t_2$*

Attempts at resolution are suggested by embodying a pronounced time dependence into the net repository entry rate $g_{+r}(t)$ and exit rate parameter $\delta_r(t)$ such that

$$\frac{dX_r}{dt} = g_{+r}(t) - \delta_r(t) X_r(t) < 0, \tag{13.7}$$

with $g_{+r}(t)$ decreasing with time and $\delta_r(t)$ now a fast variable function increasing with time. This representation constitutes a nonautonomous dynamic and embodies the concerted efforts of many specific segments of society acting in the following interests:

(a) *Device Conservation*
    Decreases in the market exit rate $g_{+r}(t)$ by extending the device operational mean-life τ_m.

(b) *Waste Management*

Increasing the remains extraction rate parameter $\delta_r(t)$ by some combination of changes in components $\rho_1(t)$, $\rho_2(t)$ and $\rho_3(t)$.

The consequences on $X_r(t)$ are suggested in Fig. 13.4(c).

13.5 Engineering: Imagination at Work

Our conceptualization of the earliest form of engineering involved Stone Age Man as the critical instigator of the primal engineering progression

$$N(t) \rightarrow E(t) \rightarrow D(t). \tag{13.8}$$

About a million years later, Contemporary Engineering continues this tradition of invention and innovation within the context of the more complex and dynamic engineering connectivity:

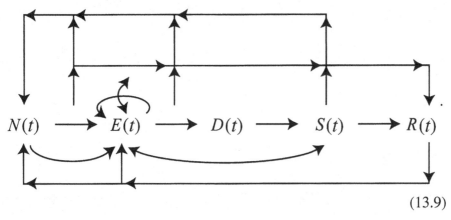

$$\tag{13.9}$$

The differences in these two symbolic depictions, (13.8) and (13.9), relate to the profound march of the engineering imagination. Not only have ingenious and useful devices been introduced but critical connections have similarly emerged; indeed, by the several arrows entering and leaving $E(t)$, engineers are evidently provided with much intellectual opportunity and process flexibility. And in this perspective of ultimate material flow closure one may identify the profound stimulation of engineering: while matter and energy transformations are intrinsically bounded, the intellectual component in the development of ingenious and useful devices — residing within $E(t)$ and $S(t)$ — is without bounds. And therefore, the conceptualization, planning, designing, developing, making, testing, implementing, improving, and disposing of devices, will evidently continue indefinitely.

13.6 To Think About

- List specific engineering challenges associated with processes $\rho_1(t)$, $\rho_2(t)$, and $\rho_3(t)$. Why are these often ignored in engineering project planning?
- Itemize selected aspects of process $\omega_1(t)$, $\omega_2(t)$, and $\omega_3(t)$ for particular classes of industries.
- Figure (13.1) provides for an explicit accounting of matter flow. Consider this pattern for an explicit accounting of energy flow.

Appendices

Appendices

Symbolic Notation

Archaeological investigations have shown that the earliest evidence of numerical concepts can be traced to the Sumerians, about 5000 years ago. These ancient people used clay tokens and tablets marked with reed-impressed symbols which signified both the kind and number of each kind of object of interest to ancient agriculturists. Thus, the idea of symbols for sets (goats, sheep, quantity of barley, ...) and the idea of symbols for numbers (1, 2, 6, 60, ...) was born and these were then combined with the operations of addition and subtraction to enable primal forms of trade and commerce to become established.

By about 2500 BCE, Babylonian scribes had discovered the concepts of multiplication and division leading to calculations for irrigation canals and, subsequently, the Egyptians used similar methods for the construction of their pyramids. Additionally, these two mathematical operations allowed the calculation of taxes thereby contributing to kingdom economies.

Present notation for the above numerical operations, $(+, -, \times, \div)$, did not appear until the mid 1500s in England.

In the late 1600s, Isaac Newton, England, and Gottfried Leibniz, Germany, independently developed two additional mathematical operations: differentiation, and integration. These two procedures and associated physical interpretations proved to be of profound importance to the establishment of science and the development of research methods. The notation now used, df/dx and $\int f dx$, is due to Leibnitz.

Table A.1 Summary of symbolic notation and primary mathematical
operations.

Universal Operation	Symbolic Notation	Initial Application
addition, subtraction	$+, -$	trade and commerce
multiplication, division	$\times, \div$	construction and taxation
differentiation, integration	$df/dx, \int f\,dx$	science and research

Three features associated with symbolic notation need to be stressed:

(a) *Various Uses*
   It is not uncommon to use the same symbol for different purposes. For
   example, the $(+, -)$ symbols are also used to indicate the electrical
   charge on an object, identify a specific domain on the number line,
   list net resource assets in statements of account, define the direction
   of a vector, etc.

(b) *Cognitive Imagery*
   Engineers often associate visual depictions with particular symbols: accumulation of stock $(+)$, area extent $(\times)$, tangent of a curve
   (df/dx), parameter inadequacy $(-)$, vector construction $(A \times B)$,
   area under a curve $(\int f\,dx)$, etc.

(c) *Inventive Extensions*
   Users of mathematics have become inventive at developing spe-
   cial symbolic notation to add efficiency and effectiveness to their
   work. Among those widely used in engineering we note the Complex
   Variable (z), Union of Sets $(A \cup B)$, Laplace Transform $(\mathcal{L}(f(x)))$,
   Dynamical Systems $(dX_i/dt, i = 1, 2, \ldots)$, Limits $(\lim_{t \to t_\infty})$,
   Operational Mapping $(f : A \to B \to C \to \cdots)$, etc.

Our use of engineering progressions in this text relates directly to
our sets $N(t)$, $E(t)$, $D(t)$, $S(t)$, and $R(t)$, to operational mappings (e.g.
$N \to E \to D \to \cdots$), to superimposed feedback (e.g. $E \to D \to S$),

and to the use of differential equations in the tradition of dynamical
systems, (dX_i/dt), with the latter supplemented by the graphical interpretation of differential equations.

Time Coordinates

Since events and processes of common interest are conveniently associated with a time axis, the identification of a suitable reference point t_0 becomes important. Interestingly, while there exists wide acceptance of the adoption of convenient natural time intervals — e.g. Δt of a day, a lunation, or a year — there exist no universal agreement on a suitable reference t_0.

The most widely used reference coordinate takes t_0 as approximately corresponding to the birth of Jesus Christ about 2000 years ago qualifying the time before/after by BC/AD. In contrast, Judaism has declared t_0 to be set at 3760 on the BC scale which dates to the calling of Abraham, while Islam measures time t_0 from 622 AD to mark the event when the Prophet Muhammad fled from Mecca to Medina. It is to be noted that some USA government documents are reference dated to $t_0 = 1776$ AD (year of the American Declaration of Independence) and some French documents use $t_0 = 1792$ AD (date of the French Revolution).

In order to minimize a specific religious or political focus while still retaining the most widely used BC/AD demarcation as a point of time reference, a practice has emerged to identify events for $t > t_0 = 0$ on the AD scale as CE (Common Era) and for $t < t_0 = 0$ on the BC scale as BCE (Before Common Era). Also, the time-axis label of Before Present (BP) is widely used when referring to a sufficiently distant past.

For purposes of simplification while still retaining familiar dates, it is here chosen to use the following increasingly common convention:

(a) For all earliest times, the designation of Before Present (BP) applies, typically from the estimated time of the Big Bang event $\sim 10^{10}$ years ago until about 10^4 years in the past.

(b) For times beginning with the Agricultural Revolution, about $\sim 10^4$ BP, and ending about 2000 years ago, the label Before Common Era (BCE) is adopted.

(c) For time coordinates beginning about 2000 years ago, the label Common Era (CE) is used.

By common practice, for time coordinate beginning about 1000 years ago (i.e. ~ 1000 AD $= 1000$ CE), all time coordinate labels are dropped.

Ancient Inventions

The primal human drive for invention and innovation has led to the establishment and subsequent evolution of an impressive range of devices already in Ancient times. Some of these ingenious devices from ~4000 BCE onward are enumerated in sequential form, also noting their apparent geographical origin:

(a) ~4000 (±600) BCE:
- sun dried bricks (Mesopotamia)
- copper smelting (Egypt)
- kilns: perforated surfaces placed over fire (Mesopotamia)
- weaving (Mesopotamia, Asia)
- woodwind and stringed musical instruments (various locations)

(b) ~3000 (±500) BCE:
- reed boats with square sails and oars (Egypt)
- stone buildings (Egypt)
- scratch plow (Mesopotamia)
- cotton growing (Asia)
- lunar calendar: ~29 days (Sumer)
- sun dial shadow stick (Egypt)
- wheel and cuneiform script (Sumer)

(c) ~2500 (±450) BCE:
- soldering of ornaments (Sumer)
- square and triangular sailships with steering rudders (Egypt)
- papyrus writing/drawing (Egypt)
- large bronze castings of urns (Asia)

313

- solar calendar: ~365 days, 12 equal periods, plus 5 days (Egypt)

(d) ~2000 (±400) BCE:
- glass container manufacture (Egypt)
- fine metal working (Egypt)
- black and red ink (China)
- gut parchment (Egypt)
- phonetic alphabet of 22 symbols (Phoenicia)

(e) ~1250 (±350) BCE:
- saddle and reins for horses (Turkey)
- bellows for iron production (Turkey, Near East)

(f) ~1000 (±300) BCE:
- leather manufacture (Sumer, Egypt)
- hard iron by hammering and quenching (Asia Minor)
- pulley (Middle East)
- suspended lodestone and magnetized iron sliver placed on cork in water bowl for direction indication (China)

(g) ~750 (±300) BCE:
- siege towers and battering ram (Assyria)
- bireme/trireme: ships with 2 or 3 banks of rowers (Greece)
- coins (Turkey)
- pyrotechnics, fireworks (China)
- torsion catapult (Middle East, Roman Empire)

(h) ~500 (±250) BCE:
- cast-iron (China)
- water clocks (Greece)

(i) ~250 (±200) BCE:
- glass making (Egypt)
- Archimedes screw (Greece)
- paper: mixed and dried vegetable fiber (China)
- block and tackle (Greece)
- block printing based on engraved stone (China)
- horse collar (China)
- crank handle (China)

(j) ~1 (±100) BCE/CE:
- water wheel for grinding wheat (Roman Empire)
- hardwood roller bearing (Europe)

- mechanical toys (Greece)
- Julian Calendar: 365 days/year with every fourth year of 366 days, 12 months (Roman Empire); the accumulated 10 day error was corrected in 1582 which together with other changes introduced the Gregorian calendar (Vatican)
- undershot water wheel (Roman Empire)

(k) ~100 (±200) CE:
- glass blowing (Middle East)
- wine press (Greece)
- glass window (Roman Empire)
- seismograph (China)

(l) ~250 (±200) CE:
- abacus (China)
- iron chains for suspension bridges (India)
- clinker boats: overlapping hull boards fastened with iron rivets (Northern Europe)

(m) ~500 (±200) CE:
- overshot water wheel (Roman Empire)
- water-powered sawmill (Northern Europe)
- porcelain, gunpowder, and matches (China)

This Ancient era, ~8000 BCE → 500 CE, is evidently characterized by an extraordinary inventiveness in the making of ingenious devices.

Cyclic Representations

Engineers are familiar with differential equations for the description of time dependent physical processes. Less clear may be the graphical representations of operational mappings and cyclic progressions. We illustrate such representations for a general case of engineering relevance involving activities for which time may be implied, explicit, or implicit.

D.1 Literal Representation: Time *Implied*

Supervising engineer E_s has noted that engineer E_x is especially adept at resolving a particular class of technical assignments or problems P. Supervisor E_s has also observed that E_x is very methodical in such problem analyses and associated reporting: E_x first becomes familiar with the problem, checks out texts, makes some personal contacts, reviews selected journal articles, and obtains specialized information from the library and the Internet, all of which provide E_x with a body of relevant information I; E_x then undertakes selected analyses A, variously integrating the available information I and checking assumptions; as feasible solutions emerge, E_x prepares a report R and also documents some peripheral background information in file F. The report R is then submitted to supervising engineer E_s and E_x thereupon is available for the next assignment, thereby repeating the cyclical pattern.

D.2 Graphical Representation: Time *Explicit*

The time explicit representation emphasizes the important feature that all activities are associated with a specific beginning time coordinate t_i and require a time interval $(t_j - t_i)$ to complete or terminate. For the above literal case, the initial problem P may have become apparent at t_0, i.e. $P(t_0)$. Supervising engineer receives it for action at t_1, i.e. $E_s(t_1)$, and following his own review assigns it to E_x at t_2, i.e. $E_x(t_2)$. Engineer E_x begins information gathering at t_3, i.e. $I(t_3)$, and undertakes specific analyses at t_4, i.e. $A(t_4)$. Eventually feasible solutions emerge with reporting beginning about t_5, i.e. $R(t_5)$, background documentation is filed at t_6, i.e. $F(t_6)$, and the report is submitted at t_7, i.e. $E_s(t_7)$. Supervisor E_s may add some changes to the report and release it at t_8, $R^0(t_8)$. With this notation, the literal representation now takes on a time explicit heterogeneous progression with time evolving from left to right:

$$I(t_3) \qquad\qquad F(t_6)$$
$$\searbow \qquad\qquad \nearrow$$
$$P(t_0) \to E_s(t_1) \to E_x(t_2) \to A(t_4) \to R(t_5) \to E_s(t_7) \to R^0(t_8).$$

$$(D.1)$$

Such representations are most useful for purposes of scheduling, billing, project progress itemization, patent priority establishment, and many other engineering requirements.

D.3 Graphical Representation: Time *Implicit*

While time implied and time explicit representations are evidently most informing about specifics of progressions, time implicit representations may be most effective in identifying cyclical process patterns and evolutionary features of progressions. In these representations, time is a continuous flow variable and time coordinates are de-emphasized. Further, the various linkages are displayed on a 2D display surface — but a 3D progression depiction is evident with time evolving perpendicularly. The above one-cycle case (D.1) would typically be displayed as follows, starting with $P(t)$:

$$R^0(t) \quad \nwarrow \nearrow \quad E_x(t) \quad \searrow \swarrow \quad I(t)$$

$$E_s(t) \qquad\qquad A(t)$$

$$\qquad\qquad\qquad \odot \; \textit{Time arrow} \qquad (D.2)$$
$$P(t) \qquad R(t) \qquad \textit{perpendicular}$$
$$\qquad\qquad\qquad \textit{upward}$$

$$F(t)$$

A helpful interpretation is to view the closure circle as a projection on a 2D plane suggesting an evolving spiral trajectory for successive projects about some perpendicular time line with linkages entering and exiting at various points on the spiral:

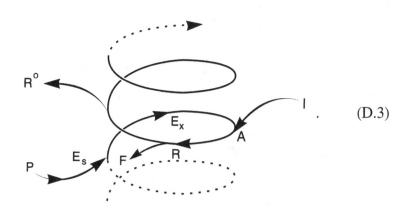

$$\text{(D.3)}$$

Some commonly encountered variations of the case here described can readily be incorporated into the time implicit representation. For example, if it is required for a draft analysis A^1 to be conveyed to the supervising engineer for a preliminary assessment, then a feedback linkage $A^1 \rightarrow E_s$ may be indicated; similarly if, additionally, E_s is expected to contribute to the preparation of the final report R — after preliminary reporting of E_x — then a feedforward linkage $E_s \rightarrow R^1$ may also be introduced. Both additions now suggest this representation as follows:

feedforward

$$P(t) \rightarrow E_s(t) \rightarrow E_x(t) \rightarrow A^1(t) \rightarrow A(t) \rightarrow R^1(t) \rightarrow R(t) \rightarrow E_s(t) \rightarrow R^0(t)$$

$$I(t_s)$$

$$F(t)$$

feedback

$$\text{(D.4)}$$

The spiral pattern about some time lines may be characterized by digressions and recursions — but still forward in time:

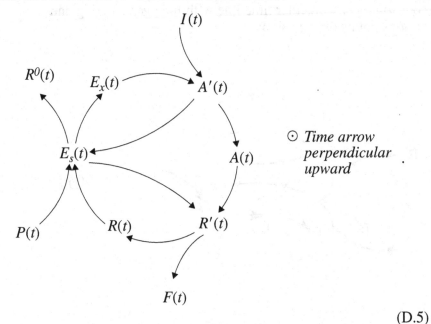

⊙ *Time arrow*
perpendicular
upward

(D.5)

Engineering practice can indeed display complex patterns in time.

Bibliography

Our objective has been to provide for a readable and organized perspective of *Engineering in Time*, free of detailed referencing which would evidently be disruptive to the reader. Many extended elaborations are readily available in the open literature with the following especially recommended:

R.M. Adams, *Paths of Fire*, Princeton University Press (1996); R.U. Ayres and L.W. Ayres, *Industrial Ecology*, Elgar Publ. (1996); G. Basalla, *Evolution of Technology*, Cambridge University Press (1988); W.E. Bijker and J. Law, Eds., *Shaping Technology — Building Society*, MIT Press (1992); D.P. Billington, *The Innovators*, Wiley & Sons (1996); A. Borgmann, *Technology and the Character of Contemporary Life*, University of Chicago, Press (1984); J. Burke, *Connections*, Little, Brown and Co. (1978); D. Cardwell, *Fontana History of Technology*, Fontana Press (1994); R.W. Clark, *Works of Man*, Century Publishing (1985); J. Dyson, *A History of Great Inventions*, Carroll & Graf (2001); J.K. Finch, *The Story of Engineering*, Anchor Books (1960); S.C. Florman, *The Civilized Engineer*, St. Martin's Press (1986); V-A. Giscard d'Estaing and M. Young, *Inventions and Discoveries*, Facts on File (1993); A. Gruebler, *Technology and Global Change*, Cambridge University Press (1998); T. Homer-Dixon, *The Ingenuity Gap*, Knopf (2000); D.S. Landes, *The Wealth and Poverty of Nations*, Norton (1998); J.H. Lienhard, *The Engines of Our Ingenuity*, Oxford University Press (2000); J.E. McClellan and H. Dorn, *Science and Technology in World History*, Johns Hopkins University Press (1999); D.C. Mowery and N. Rosenberg, *Paths of Innovation*, Cambridge University Press (1998); A. Pacey, *The Maze of Ingenuity*, MIT Press (1992); H. Petroski, *The Evolution of Useful Things*, Vintage Books (1994); R. Pool, *Beyond Engineering*, Oxford University Press (1997); J. Rae and R. Volti, *The Engineer in History*, Lang Publ. (1993); G.P. Richardson, *Feedback Thought in Social Science and Systems Theory*, University of Pennsylvania Press (1991); W. Rybczynski, *Taming the Tiger*, Penguin (1985); H.A. Simon, *Sciences of the Artificial*, MIT Press (1996); V. Smil, *Energies*, MIT Press (1999); D.L. Spar, *Ruling the Waves*, Harcourt (2001); T.I. Williams, *History of Invention*, Facts on File (1987).

Index

About the Authors

Archie Harms, Professor Emeritus of Engineering Physics at McMaster University, Hamilton, ON, holds degrees from the University of British Columbia and the University of Washington. He has authored or co-authored over 150 papers in refereed journals as well as four books. A professional engineer, his research and industrial experience includes water resources, engineering systems theory, and emerging nuclear energy concepts.

Brian Baetz is Professor of Civil and Environmental Engineering at Tulane University, New Orleans, LA. He has received engineering degrees from the University of Toronto and Duke University and is registered as a professional engineer. His research is in the areas of municipal solid waste management systems and infrastructure planning for sustainable communities.

Rudi Volti, Professor of Sociology, Pitzer College, Claremont, CA, is a graduate of the University of California and Rice University. A former Director of the Claremont College Program on Science, Technology, and Society, he has published extensively on issues of technology and society. He is coauthor of *The Engineer in History* (Lang, 1993, 2001).

Printed in the United States
By Bookmasters